Joseph's 487 questions to Precalculus

Joseph Jones

ISBN 978-1-300-57354-8

TABLE OF CONTENTS

Chapter 1   Function and inverse function

▶ We determine whether a function is odd, even, or neither. To do so, we substitute $-X$ into X and look at the result. If $f(-X)=-f(X)$, the function is odd, which means it is rotated $180°$ about the origin. If $f(-X)=f(X)$, the function is even; this implies that the function is symmetric about the y-axis. If $f(-X)$ does not equal either $-f(X)$ or $f(X)$, the function is neither.

Example 1:
Determine whether the expression is an odd function, even function, or neither.

$$f(X)=-2X^4+5X^2-2X+3 \qquad \text{We substitute } -X.$$

$$f(-X)=-2(-X)^4+5(-X)^2-2(-X)+3$$
$$=-2X^4+5X^2+2X+3$$
$$=-(2X^4-5X^2-2X-3)$$

$$f(-X)\neq-f(X)$$
$$f(-X)\neq f(X)$$

Neither

The function is neither even nor odd.

Example 2:
Determine whether the expression is an odd function, even function, or neither.

$$f(X)=4X^5+3X^3+X \qquad \text{We substitute } -X.$$

$$f(-X)=4(-X)^5+3(-X)^3+(-X)$$
$$=-4X^5-3X^3-X$$
$$=-(4X^5+3X^3+X)$$

$$f(-X)=-f(X)$$

Odd function

We have $f(-X)=-f(X)$, so we have an odd function.

» What goes between the parenthesis of the function $f(X)$ will be substituted into the X of that function equation.

$$f(X) = Y$$
$$\downarrow \quad \swarrow$$
$$(X, Y)$$

First, function notation looks like $f(X) = Y$. What goes between the parenthesis is the x-value. $f(X)$ is the y-value.

Example 3:
Evaluate.

$$f(X) = 2^{2X+1} + 7$$
$$f(2) =$$

$$f(2) = 2^{2(2)+1} + 7$$

What goes between the parenthesis of the function will be substituted into X of that equation.

$$= 2^5 + 7 = 39$$

Example 4:
Evaluate.

$$f(X) = -5e^{2X}$$
$$f(5) =$$

$$f(5) = -5e^{2(5)}$$

What goes between the parenthesis of the function will be substituted into X of that equation.

$$= -5e^{10} \approx -110,132$$

» We apply the horizontal-line test to determine whether the inverse graph will be a function. We scan the horizontal line and, if the graph intersects the horizontal line at most once, the graph passes the horizontal-line test and therefore has an inverse that is a function. The graph that passes the horizontal-line test is one-to-one. If the graph does not pass the horizontal-line test, the inverse is not a function.

Example 5:

Determine whether an inverse function exists.

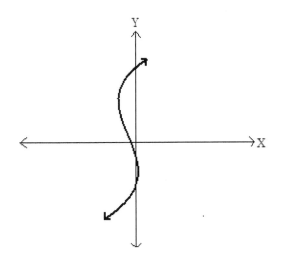

We apply the horizontal-line test.

The graph passes horizontal-line test. When we scan a horizontal line up and down, the line intersects the graph at most once. Because the graph passes the horizontal-line test, inverse function exists.

» We will study about function notations. We focus on the addition, subtraction, multiplication, and division.

The function may be notated with different letters, but they are essentially the same.

$$f(X) = Y \qquad\qquad g(X) = Y$$
$$p(X) = Y \qquad\qquad h(X) = Y$$

We have notations for addition, subtraction, multiplication, and division.

Adding two function equations	$(A + B)(X) = A(X) + B(X)$
Subtracting two function equations	$(A - B)(X) = A(X) - B(X)$
Multiplying two function equations	$(AB)(X) = A(X)B(X)$
Dividing two function equations	$\left(\dfrac{A}{B}\right)(X) = \dfrac{A(X)}{B(X)}$

Example 6:

$$A(X) = X^8 - X^5 + 3X^2 + 7$$
$$B(X) = X^4 + 5X^2 - 1$$

We have two function equations, $A(X)$ and $B(X)$, given. We are subtracting these two equations.

$$(A-B)(X)=$$

$$(A-B)(X)= A(X)- B(X)$$
$$= (X^8 - X^5 + 3X^2 + 7)- (X^4 + 5X^2 - 1)$$
$$= X^8 - X^5 + 3X^2 + 7 - X^4 - 5X^2 + 1$$
$$= X^8 - X^5 - X^4 - 2X^2 + 8$$

We combine the like terms.

Example 7:

$$A(X)= (X^2 - 10)$$
$$B(X)= (X + 21)$$

We multiply the two function equations.

$$(AB)(X)=$$

$$(AB)(X)= A(X)B(X)$$
$$= (X^2 - 10)(X + 21)$$
$$= X^3 + 21X^2 - 10X - 210$$

We substitute function equations. We multiply each term in one parenthesis by each term in the other parenthesis. We then combine the like terms.

Example 8:

$$A(X)= 2X^2 - 9X - 5$$
$$g(X)= X - 5$$

We divide the two function equations.

$$\left(\frac{A}{B}\right)(X)=$$

$$\left(\frac{A}{B}\right)(X)= \frac{A(X)}{B(X)}$$

We substitute the function equations.

$$= \frac{2X^2 - 9X - 5}{X - 5}$$

We factor the numerator.

$$= \frac{(2X + 1)(X - 5)}{X - 5}$$

$X - 5$ in the numerator and in the denominator cancels out.

$$= 2X + 1$$

» Inverse function is a function reflected across $Y = X$.

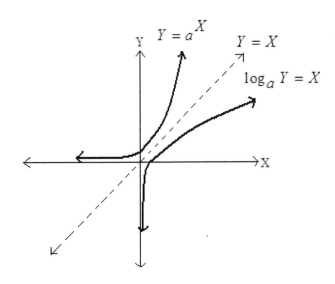

$\log_a Y = X$ is an inverse function of the exponential function $Y = a^X$.

$\log_a Y = X$ is reflected across the $Y = X$.

» Not all graphs have inverses that are functions.

» We studied that an inverse is reflected across the graph $Y = X$. Another interpretation is that the x-coordinate and y-coordinate are switched in the inverse graphs.

$f(X)$ $f^{-1}(X) \longrightarrow$ inverse of $f(X)$

X	Y
-2	-3
-1	2
0	7
1	12
2	17

X	Y
-3	-2
2	-1
7	0
12	1
17	2

We have some coordinate points of the graph $f(X)$. The coordinates of $f^{-1}(X)$ have x-coordinates and y-coordinates of $f(X)$ switched.

» To obtain an inverse function equation, we switch the places of X and Y, and then, solve for Y.

$$Y = 5X + 7$$

To find inverse, we switch the X and Y.

$$X = 5Y + 7$$
$$X - 7 = 5Y$$
$$Y = \frac{X - 7}{5}$$

Then, we try to get Y by itself.

$Y = \dfrac{X - 7}{5}$ is an inverse of $Y = 5X + 7$.

Example 9:
Find the inverse function.

$$D(X) = 8(X - 5) + 7$$
$$Y = 8(X - 5) + 7$$

We replace $D(X)$ with Y because $D(X) = Y$.

$$X = 8(Y - 5) + 7$$

We switch the X and Y.

$$X - 7 = 8(Y - 5)$$
$$X - 7 = 8Y - 40$$
$$8Y = X + 33$$
$$Y = \frac{X + 33}{8}$$

We get Y by itself on one side.

$$D^{-1}(X) = \frac{X + 33}{8}$$

We can replace the Y with an inverse function sign, $D^{-1}(X)$.

Chapter 1   Function and inverse function

1. Determine whether the function is even, odd, or neither. $f(X) = X^3 - 2X^2 + 2$	2. Determine whether the function is even, odd, or neither. $f(X) = X^4 - 3X^2 + 1$
3. Determine whether the function is even, odd, or neither. $f(X) = 12X^7 - 5X^5 + 3X^3 + 11X$	4. Determine whether the function is even, odd, or neither. $f(X) = -X^9 - 2X^7 + X^5 - X^3 + 7X$
5. Determine whether the function is even, odd, or neither. $f(X) = 5X^7 + 2X^4 - 2X + 1$	6. Determine whether the function is even, odd, or neither. $f(X) = 5X^8 + 3X^6 + 5X^4 + X^2 - 4$
7. Evaluate. $f(X) = 3^{X+1} + 1$ $f(1) =$	8. Evaluate. $f(X) = 5^{X-1} + 3$ $f(3) =$
9. Evaluate. $f(X) = X^2 - 5X + 1$ $f(-2) =$	10. Evaluate. $f(X) = (X - 4)^2 + 3X + 7$ $f(5) =$

Chapter 1   Function and inverse function

11. Evaluate. $f(X) = X^5 - 4X^2 + X - 1$ $f(1) =$	12. Evaluate. $f(X) = 2X^4 - 5X^3 + X - 11$ $f(-2) =$
13. Apply horizontal-line test to determine whether an inverse function exists. 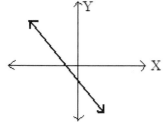	14. Apply horizontal-line test to determine whether an inverse function exists. 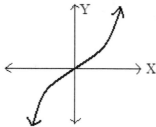
15. Apply horizontal-line test to determine whether an inverse function exists. 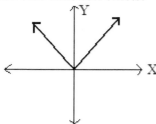	16. Apply horizontal-line test to determine whether an inverse function exists. 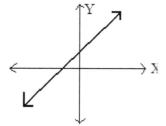
17. Evaluate. $f(X) = \dfrac{3}{1+X} - 2$ $f(2) =$	18. Evaluate. $f(X) = \dfrac{5}{X+4} + 7$ $f(1) =$
19. Evaluate. $f(X) = \dfrac{10}{X-2} + 3$ $f(7) =$	20. Evaluate. $f(X) = \dfrac{30}{X-1} + 11$ $f(6) =$

Chapter 1   Function and inverse function

21. Evaluate. $f(X) = 5e^{2X}$ $f(5) =$	22. Evaluate. $f(X) = Xe^{3(4)}$ $f(2) =$
23. Evaluate. $f(X) = 10e^{-2X}$ $f(1) =$	24. Evaluate. $f(X) = 4e^{-5X}$ $f(3) =$
25. Evaluate. $f(X) = 3e^{2X}$ $f(1) =$	26. Evaluate. $f(X) = 5e^{-2X}$ $f(5) =$
27. Evaluate. $f(X) = Xe^{-2(2)}$ $f(1) =$	28. Evaluate. $f(X) = -4e^{-2X}$ $f(7) =$
29. $A(X) = X + 2$ $B(X) = 2X + 3$ $(A + B)(X) =$	30. $A(X) = 2X - 5$ $B(X) = 7X$ $(A - B)(X) =$

Chapter 1   Function and inverse function

31. $$A(X) = X + 14$$ $$B(X) = 2X - 1$$ $$(AB)(X) =$$	32. $$A(X) = X^3 + 2X + 1$$ $$B(X) = -4X^2 + 3X - 1$$ $$(A + B)(X) =$$
33. $$A(X) = X^2 + 16X + 48$$ $$B(X) = X + 4$$ $$\left(\frac{A}{B}\right)(X) =$$	34. $$A(X) = 2X - 1$$ $$B(X) = 3X^2 + 4X + 5$$ $$(AB)(X) =$$
35. $$A(X) = X^3 - 5X^2 + X - 1$$ $$B(X) = X^2 + 4X - 5$$ $$(A - B)(X) =$$	36. Find the inverse function. $$D(X) = 2X + 17$$ $$D^{-1}(X) =$$
37. Find the inverse function. $$D(X) = 4(X + 5) + 2$$ $$D^{-1}(X) =$$	38. Find the inverse function. $$D(X) = 2(X^3 - 4)$$ $$D^{-1}(X) =$$
39. Find the inverse function. $$D(X) = 5(2X - 1) + 10$$ $$D^{-1}(X) =$$	40. Find the inverse function. $$D(X) = 3(X^3 - 2) + 15$$ $$D^{-1}(X) =$$

Chapter 1 Function and inverse function
Answer key

1. $f(X) = X^3 - 2X^2 + 2$
   $f(-X) = (-X)^3 - 2(-X)^2 + 2$
   $= -X^3 - 2X^2 + 2$
   Neither

2. $f(X) = X^4 - 3X^2 + 1$
   $f(-X) = (-X)^4 - 3(-X)^2 + 1$
   $= X^4 - 3X^2 + 1$
   $f(-X) = f(X)$
   Even function

3. $f(X) = 12X^7 - 5X^5 + 3X^3 + 11X$
   $f(-X) = 12(-X)^7 - 5(-X)^5 + 3(-X)^3 + 11(-X)$
   $= -12X^7 + 5X^5 - 3X^3 - 11X$
   $= -(12X^7 - 5X^5 + 3X^3 + 11X)$
   $f(-X) = -f(X)$
   Odd function

4. $f(X) = -X^9 - 2X^7 + X^5 - X^3 + 7X$
   $f(-X) = -(-X)^9 - 2(-X)^7 + (-X)^5 - (-X)^3 + 7(-X)$
   $= X^9 + 2X^7 - X^5 + X^3 - 7X$
   $= -(-X^9 - 2X^7 + X^5 - X^3 + 7X)$
   $f(-X) = -f(X)$
   Odd function

5. $f(X) = 5X^7 + 2X^4 - 2X + 1$
   $f(-X) = 5(-X)^7 + 2(-X)^4 - 2(-X) + 1$
   $= -5X^7 + 2X^4 + 2X + 1$
   Neither

6. $f(X) = 5X^8 + 3X^6 + 5X^4 + X^2 - 4$
   $f(-X) = 5(-X)^8 + 3(-X)^6 + 5(-X)^4 + (-X)^2 - 4$
   $= 5X^8 + 3X^6 + 5X^4 + X^2 - 4$
   $f(-X) = f(X)$
   Even function

7. $f(X) = 3^{X+1} + 1$
   $f(1) = 3^{1+1} + 1$
   $= 3^2 + 1$
   $= 10$

8. $f(X) = 5^{X-1} + 3$
   $f(3) = 5^{3-1} + 3$
   $= 5^2 + 3$
   $= 28$

9. $f(X) = X^2 - 5X + 1$
   $f(-2) = (-2)^2 - 5(-2) + 1$
   $= 4 + 10 + 1$
   $= 15$

10. $f(X) = (X - 4)^2 + 3X + 7$
    $f(5) = (5 - 4)^2 + 3(5) + 7$
    $= 1 + 15 + 7$
    $= 23$

11. $f(X) = X^5 - 4X^2 + X - 1$
    $f(1) = (1)^5 - 4(1)^2 + (1) - 1$
    $= 1 - 4 + 1 - 1$
    $= -3$

12. $f(X) = 2X^4 - 5X^3 + X - 11$
    $f(-2) = 2(-2)^4 - 5(-2)^3 + (-2) - 11$
    $= 2(16) - 5(-8) + (-2) - 11$
    $= 32 + 40 - 2 - 11$
    $= 59$

13. The graph passes horizontal-line test; inverse function exists.

14. The graph passes horizontal-line test; inverse function exists.

15. No inverse function

16. The graph passes the horizontal-line test; inverse function exists.

17. $f(X) = \dfrac{3}{1+X} - 2$

$f(2) = \dfrac{3}{1+2} - 2$

$= 1 - 2$

$= -1$

18. $f(X) = \dfrac{5}{X+4} + 7$

$f(1) = \dfrac{5}{1+4} + 7$

$= \dfrac{5}{5} + 7$

$= 8$

19. $f(X) = \dfrac{10}{X-2} + 3$

$f(7) = \dfrac{10}{7-2} + 3$

$= 2 + 3$

$= 5$

20. $f(X) = \dfrac{30}{X-1} + 11$

$f(6) = \dfrac{30}{6-1} + 11$

$= 6 + 11$

$= 17$

21. $f(X) = 5e^{2X}$

$f(5) = 5e^{2(5)}$

$= 5e^{10}$

$= 110{,}132.33$

22. $f(X) = Xe^{3(4)}$

$f(2) = 2e^{12}$

$= 325{,}509.58$

23. $f(X) = 10e^{-2X}$

$f(1) = 10e^{(-2)(1)}$

$= 10e^{-2}$

$= \dfrac{10}{e^2}$

≈ 1.353

24. $f(X) = 4e^{-5X}$

$f(3) = 4e^{(-5)(3)}$

$= 4e^{-15}$

$= \dfrac{4}{e^{15}}$

$= 0.0000012$

25. $f(X) = 3e^{2X}$

$f(1) = 3e^{(2)(1)}$

$= 3e^2$

$= 22.17$

26. $f(X) = 5e^{-2X}$

$f(5) = 5e^{(-2)(5)}$

$= 5e^{-10}$

$= \dfrac{5}{e^{10}}$

$= 0.000227$

27. $f(X) = Xe^{-2(2)}$

$f(1) = e^{-4} = \dfrac{1}{e^4}$

$= 0.018316$

28. $f(X) = -4e^{-2X}$

$f(7) = -4e^{(-2)(7)}$

$= -4e^{-14} = \dfrac{-4}{e^{14}}$

$= -0.000003$

29. $(A+B)(X)= A(X)+ B(X)=$
    $(X+2)+(2X+3)= 3X+5$

30. $(A-B)(X)= A(X)- B(X)=$
    $(2X-5)-(7X)= 2X-5-7X =$
    $-5X-5$

31. $(AB)(X)= A(X)B(X)=$
    $(X+14)(2X-1)=$
    $2X^2 - X + 28X - 14 =$
    $2X^2 + 27X - 14$

32. $(A+B)(X)= A(X)+ B(X)=$
    $(X^3 + 2X + 1)+(-4X^2 + 3X - 1)=$
    $X^3 + 2X + 1 - 4X^2 + 3X - 1 =$
    $X^3 - 4X^2 + 5X$

33. $\left(\dfrac{A}{B}\right)(X)= \dfrac{A(X)}{B(X)} =$
    $\dfrac{X^2 + 16X + 48}{X+4} = \dfrac{(X+12)(X+4)}{X+4}$
    $= X + 12$

34. $(AB)(X)= A(X)B(X)=$
    $(2X-1)(3X^2 + 4X + 5)=$
    $6X^3 + 8X^2 + 10X - 3X^2 - 4X - 5 =$
    $6X^3 + 5X^2 + 6X - 5$

35. $(A-B)(X)= A(X)- B(X)=$
    $(X^3 - 5X^2 + X - 1)-(X^2 + 4X - 5)=$
    $X^3 - 5X^2 + X - 1 - X^2 - 4X + 5 =$
    $X^3 - 6X^2 - 3X + 4$

36. $Y = 2X + 17$
    $X = 2Y + 17$
    $Y = \dfrac{X-17}{2}$
    $D^{-1}(X)= \dfrac{X-17}{2}$

37. $Y = 4(X+5)+2$
    $X = 4(Y+5)+2$
    $\dfrac{X-2}{4} = Y + 5$
    $D^{-1}(X)= \dfrac{X-2}{4} - 5$

38. $Y = 2(X^3 - 4)$
    $X = 2(Y^3 - 4)$
    $X = 2Y^3 - 8$
    $\sqrt[3]{\dfrac{X+8}{2}} = Y$
    $D^{-1}(X)= \sqrt[3]{\dfrac{X+8}{2}}$

39. $Y = 5(2X-1)+10$
    $X = 5(2Y-1)+10$
    $X = 10Y + 5$
    $\dfrac{X-5}{10} = Y$
    $D^{-1}(X)= \dfrac{X-5}{10}$

40. $Y = 3(X^3 - 2)+15$
    $X = 3(Y^3 - 2)+15$
    $X = 3Y^3 + 9$
    $\sqrt[3]{\dfrac{X-9}{3}} = Y$
    $D^{-1}(X)= \sqrt[3]{\dfrac{X-9}{3}}$

Chapter 2   Logarithms

▶ Logarithm is an inverse function of an exponential function.

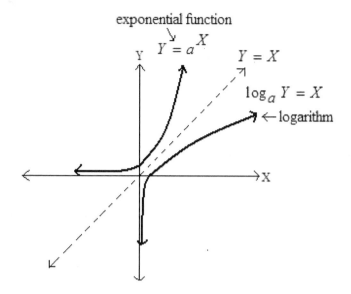

We can express an exponential function into logarithm or logarithm into an exponential function.

$$\log_A C = B \qquad \leftrightarrow \qquad A^B = C$$
$$A^B = C \qquad \leftrightarrow \qquad \log_A C = B$$

$$\log_{base} value = exponent \qquad \longleftrightarrow \qquad (base)^{exponent} = value$$

$$(base)^{exponent} = value \qquad \longleftrightarrow \qquad \log_{base} value = exponent$$

Example 1:
Express the logarithm as
an exponential function.

$$\log_F C = H$$

We can express logarithm as an exponential function. $\log_F C = H \leftrightarrow F^H = C$

$$F^H = C$$

» We can solve for the value of X by converting logarithm to an exponential function.

Example 2:
Solve for X without calculator.

$$\log_2 X = 7$$
We convert the logarithm to an exponential function.

$$2^7 = X$$
We think of 2 to the 7th power equals what number.

$$X = 128$$

» We will study logarithm rule in expanding the logarithms.

$$\log X^a = a \log X$$
With logarithm, we can bring down the exponent in front of "log" sign.

$$\log(XY) = \log X + \log Y$$

$$\log\left(\frac{X}{Y}\right) = \log X - \log Y$$

Example 3:
Expand the logarithms.

$$\log\left(X^2 Y^3 Z\right) =$$

$$\log X^2 + \log Y^3 + \log Z =$$
We apply the rule $\log(XY) = \log X + \log Y$.

$$2\log X + 3\log Y + \log Z$$
We can apply the rule $\log X^a = a \log X$.

Example 4:
Expand the logarithms.

$$\log\left(\frac{X^2}{Y^5 Z^3}\right) =$$
We apply the logarithm rules.

$$\log X^2 - \log Y^5 - \log Z^3 =$$

$$2\log X - 5\log Y - 3\log Z$$

» We have logarithm rules in combining. This is the same as expanding but flows backward.

$$a \log X = \log X^a$$

$$\log X + \log Y = \log(XY)$$

$$\log X - \log Y = \log\left(\frac{X}{Y}\right)$$

Example 5:
Combine the logarithms.

$$8 \log_3 X - 2 \log_3 Y =$$

$$\log_3 X^8 - \log_3 Y^2 = \qquad \text{We apply the logarithm rule: } a \log X = \log X^a.$$

$$\log_3 \frac{X^8}{Y^2} \qquad \text{We apply the logarithm rule: } \log X - \log Y = \log\left(\frac{X}{Y}\right).$$

» We can apply the logarithm rules to solve for the value of X in the following examples.

Example 6:
Applying logarithms, solve for X.

$$12^X = 60$$

$$\log 12^X = \log 60 \qquad \text{We take the logarithms on both sides.}$$

$$X \log 12 = \log 60 \qquad \text{We apply the logarithm rule and solve for X.}$$

$$X = \frac{\log 60}{\log 12}$$

$$X \approx 1.6477$$

Example 7:
Applying logarithms, solve for X.

$$20^{X+5} = 52$$

$$\log 20^{X+5} = \log 52 \qquad \text{We take the logarithms on both sides.}$$

$$(X+5)(\log 20) = \log 52 \qquad \text{We apply the logarithm rule.}$$

$$X \log 20 + 5 \log 20 = \log 52$$

$$X \log 20 = \log 52 - 5 \log 20$$

$$X = \frac{\log 52 - 5 \log 20}{\log 20}$$

$$X \approx -3.681$$

Chapter 2   Logarithms

1. Express the logarithm as an exponential function. $\log_A D = B$	2. Express the logarithm as an exponential function. $\log X = Y$
3. Express the logarithm as an exponential function. $\log_X 32 = 5$	4. Express the logarithm as an exponential function. $\log_5 25 = X$
5. Express the logarithm as an exponential function. $\log_B G = D$	6. Solve for X without calculator. $\log_X 125 = 3$
7. Solve for X without calculator. $\log_7 343 = X$	8. Solve for X without calculator. $\log_4 1024 = X$
9. Solve for X without calculator. $\log_{11} X = 3$	10. Solve for X without calculator. $\log_5 X = 5$

Chapter 2   Logarithms

11. Expand the logarithms. $$\log_5\left(\frac{X^2}{Y}\right) =$$	12. Expand the logarithms. $$\log_7\left(\frac{XW}{Y^2}\right) =$$
13. Expand the logarithms. $$\log_8\left(X^2Y^3Z^2\right) =$$	14. Expand the logarithms. $$\log_7\left(\frac{X^7Y^2}{Z^2}\right) =$$
15. Expand the logarithms. $$\log_5\left(\frac{X^{\frac{1}{2}}}{Y^3Z^5}\right) =$$	16. Combine the logarithms. $$\log_7 X + 2\log_7 Y$$
17. Combine the logarithms. $$5\log_3 X - 7\log_3 Z$$	18. Combine the logarithms. $$2\log_2 X + \log_2 Y + 5\log_2 Z$$
19. Combine the logarithms. $$\frac{1}{3}\log_8 X - 3\log_8 Y - 4\log_8 Z$$	20. Combine the logarithms. $$3\log_3 X - \frac{1}{2}\log_3 Y + 5\log_3 Z$$

Chapter 2   Logarithms

21. Applying logarithms, solve for X. $2^{X-5} = 60$	22. Applying logarithms, solve for X. $3^{X+2} = 25$
23. Applying logarithms, solve for X. $5^{X-1} = 41$	24. Applying logarithms, solve for X. $3^X = 82$
25. Applying logarithms, solve for X. $5^{2X+1} = 23$	26. Applying logarithms, solve for X. $3^{2X-5} = 11$
27. Applying logarithms, solve for X. $2^{X-2} = 5^X$	28. Applying logarithms, solve for X. $3^{X+5} = 2^X$
29. Applying logarithms, solve for X. $4^{X+1} = 5^X$	30. Applying logarithms, solve for X. $2^{X-1} = 7^{X+2}$

Chapter 2 Logarithms
Answer key

1. $A^B = D$

2. $10^Y = X$

3. $X^5 = 32$

4. $5^X = 25$

5. $B^D = G$

6. $X^3 = 125$
   $X = \sqrt[3]{125}$
   $X = 5$

7. $7^X = 343$
   $7^X = 7^3$
   $X = 3$

8. $4^X = 1024$
   $4^X = 4^5$
   $X = 5$

9. $11^3 = X$
   $X = 1331$

10. $5^5 = X$
    $X = 3125$

11. $\log_5\left(\dfrac{X^2}{Y}\right) = \log_5 X^2 - \log_5 Y$
    $= 2\log_5 X - \log_5 Y$

12. $\log_7\left(\dfrac{XW}{Y^2}\right) =$
    $\log_7 X + \log_7 W - \log_7 Y^2 =$
    $\log_7 X + \log_7 W - 2\log_7 Y$

13. $\log_8\left(X^2 Y^3 Z^2\right)=$
    $\log_8 X^2 + \log_8 Y^3 + \log_8 Z^2 =$
    $2\log_8 X + 3\log_8 Y + 2\log_8 Z$

14. $\log_7\left(\dfrac{X^7 Y^2}{Z^2}\right) =$
    $\log_7 X^7 + \log_7 Y^2 - \log_7 Z^2 =$
    $7\log_7 X + 2\log_7 Y - 2\log_7 Z$

15. $\log_5\left(\dfrac{X^{\frac{1}{2}}}{Y^3 Z^5}\right) =$
    $\log_5 X^{\frac{1}{2}} - \log_5 Y^3 - \log_3 Z^5 =$
    $\dfrac{1}{2}\log_5 X - 3\log_5 Y - 5\log_3 Z$

16. $\log_7 X + 2\log_7 Y =$
    $\log_7 X + \log_7 Y^2 =$
    $\log_7 XY^2$

17. $5\log_3 X - 7\log_3 Z =$
    $\log_3 X^5 - \log_3 Z^7 =$
    $\log_3 \dfrac{X^5}{Z^7}$

18. $2\log_2 X + \log_2 Y + 5\log_2 Z =$
    $\log_2 X^2 + \log_2 Y + \log_2 Z^5 =$
    $\log_2\left(X^2 Y Z^5\right)$

19. $\dfrac{1}{3}\log_8 X - 3\log_8 Y - 4\log_8 Z =$
    $\log_8 X^{\frac{1}{3}} - \log_8 Y^3 - \log_8 Z^4 =$
    $\log_8\left(\dfrac{\sqrt[3]{X}}{Y^3 Z^4}\right)$

20. $3\log_3 X - \dfrac{1}{2}\log_3 Y + 5\log_3 Z =$

$\log_3 X^3 - \log_3 Y^{\frac{1}{2}} + \log_3 Z^5 =$

$\log_3 \dfrac{X^3 Z^5}{\sqrt{Y}}$

21. $2^{X-5} = 60$

$\log 2^{X-5} = \log 60$

$(X-5)(\log 2) = \log 60$

$X = \dfrac{\log 60}{\log 2} + 5 \approx 10.9$

22. $3^{X+2} = 25$

$\log 3^{X+2} = \log 25$

$(X+2)(\log 3) = \log 25$

$X = \dfrac{\log 25}{\log 3} - 2 \approx 0.93$

23. $5^{X-1} = 41$

$\log 5^{X-1} = \log 41$

$(X-1)(\log 5) = \log 41$

$X = \dfrac{\log 41}{\log 5} + 1 \approx 3.3$

24. $3^X = 82$

$\log 3^X = \log 82$

$X \log 3 = \log 82$

$X = \dfrac{\log 82}{\log 3}$

$X \approx 4.01$

25. $5^{2X+1} = 23$

$\log 5^{2X+1} = \log 23$

$(2X+1)(\log 5) = \log 23$

$X = \dfrac{\log 23}{2\log 5} - \dfrac{1}{2}$

$X \approx 0.474$

26. $3^{2X-5} = 11$

$\log 3^{2X-5} = \log 11$

$(2X-5)(\log 3) = \log 11$

$X = \dfrac{\log 11}{2\log 3} + \dfrac{5}{2} \approx 3.59$

27. $2^{X-2} = 5^X$

$\log 2^{X-2} = \log 5^X$

$(X-2)(\log 2) = X \log 5$

$X \log 2 - 2\log 2 = X \log 5$

$X \log 2 - X \log 5 = 2\log 2$

$X = \dfrac{2\log 2}{\log 2 - \log 5} \approx -1.51$

28. $3^{X+5} = 2^X$

$\log 3^{X+5} = \log 2^X$

$(X+5)(\log 3) = X(\log 2)$

$X \log 3 + 5\log 3 = X \log 2$

$X \log 3 - X \log 2 = -5\log 3$

$X = \dfrac{-5\log 3}{\log 3 - \log 2} \approx -13.55$

29. $4^{X+1} = 5^X$

$\log 4^{X+1} = \log 5^X$

$(X+1)(\log 4) = X \log 5$

$X \log 4 + \log 4 = X \log 5$

$X \log 4 - X \log 5 = -\log 4$

$X = \dfrac{-\log 4}{\log 4 - \log 5} \approx 6.21$

30. $2^{X-1} = 7^{X+2}$

$\log 2^{X-1} = \log 7^{X+2}$

$(X-1)(\log 2) = (X+2)(\log 7)$

$X \log 2 - \log 2 = X \log 7 + 2\log 7$

$X \log 2 - X \log 7 = 2\log 7 + \log 2$

$X = \dfrac{2\log 7 + \log 2}{\log 2 - \log 7} \approx -3.66$

Chapter 3   Conic sections

▶ The conic section includes parabola, circle, ellipse, and hyperbola.

We have an equation for the parabola: $(X - h)^2 = 4(p)(Y - k)$

p > 0

If P > 0, the parabola opens upward.

If P < 0, the parabola opens downward.

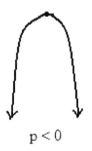

p < 0

» Vertex is at the minimum or maximum of the parabola. Based on the equation $(X - h)^2 = 4(p)(Y - k)$, the vertex is (h, k).

Vertex

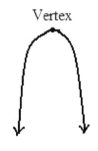

Vertex = (h, k)

Vertex

» Focus is located inside the parabola. Directrix is a line opposite the focus and the same distance from the vertex as is the distance from focus to vertex.

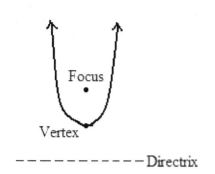
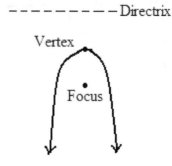

Any point on the parabola will have equal distance from the focus and equal distance from the directrix in a perpendicular line.

» Based on the vertex, the focus of the parabola opening upward or downward is $(h, k + p)$ and the directrix is $Y = k - p$.

Parabola can open side ways. The equation for parabola opening sideways is $(Y - k)^2 = 4(p)(X - h)$

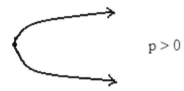

$p > 0$

When p > 0, the parabola opens toward right.

$p < 0$

When p < 0, the parabola opens toward left.

» The vertex is (h, k) based on the formula $(Y - k)^2 = 4(p)(X - h)$.

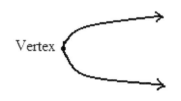   and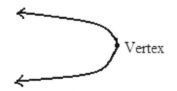

Vertex: (h, k)

» Focus is located inside the parabola. In parabola opening sideways, the directrix is a vertical line equidistant from the vertex as the focus is from the vertex.

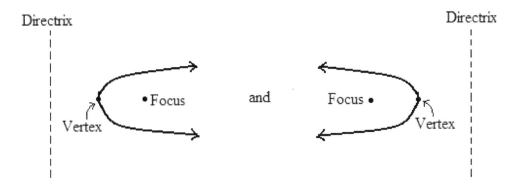

Based on the vertex (h,k), the focus of the parabola opening sideways is $(h+p,k)$ and the directrix is $X = h - p$.

Example 1:
Find the vertex, focus, and directrix.

$$(X-15)^2 = -2(Y+3)$$ We apply the parabola formula.

$$(X-h)^2 = 4(p)(Y-k)$$ We find the values of h, k and p.

$$(X-15)^2 = 4\left(-\frac{1}{2}\right)(Y-(-3))$$

$$h = 15, \ k = -3, \ p = -\frac{1}{2}$$

Vertex: $(h,k) = (15,-3)$ Vertex is (h,k).

Focus: $(h, k+p) = \left(15, -3 -\frac{1}{2}\right) = \left(15, -\frac{7}{2}\right)$ Focus for parabola opening up or down is $(h, k+p)$.

Directrix: $Y = k - p = -3 - \left(-\frac{1}{2}\right) = -\frac{5}{2}$ Directrix is $Y = k - p$.

$$Y = -\frac{5}{2}$$

Example 2:
Find the vertex, focus, and directrix.

$$(X-8) = \frac{(Y-5)^2}{12}$$ We apply the parabola formula.

$$(Y-k)^2 = 4(p)(X-h)$$ We identify the values of h, k, and p.

28

$$(Y-5)^2 = 4(3)(X-8)$$

Vertex: $(h,k) = (8,5)$ We find vertex (h,k).

Focus: $(h+p,k) = (11,5)$ Focus is $(h+p,k)$

Directrix: $X = h - p = 8 - 3 = 5$ Directrix is $X = h - p$

$$X = 5$$

» We can think about circle in an equation: $(X-h)^2 + (Y-k)^2 = r^2$. The center is (h,k). The radius is r.

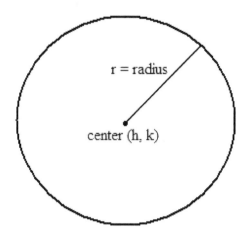

r = radius

center (h, k)

Example 3:

Find the center and radius of the circle.

$$(X+11)^2 + \left(Y - \sqrt{2}\right)^2 = 100$$

$(X-h)^2 + (Y-k)^2 = r^2$ We apply the circle formula.

$(X-(-11))^2 + \left(Y - \sqrt{2}\right)^2 = 10^2$ We identify the values of h, k, and p.

Center: $(h,k) = \left(-11, \sqrt{2}\right)$

radius $= r = 10$

» We may not always have the circle equation $(X-h)^2 + (Y-k)^2 = r^2$. We will refer to the equation $(X-h)^2 + (Y-k)^2 = r^2$ as the standard equation of circle. Sometimes, we have to convert a circle equation to $(X-h)^2 + (Y-k)^2 = r^2$ by completing the square.

Example 4:
Write the standard equation of circle
by completing the square.

$$X^2 - 40X + Y^2 + 24Y + 495 = 0$$

$$\left(X^2 - 40X\right) + \left(Y^2 + 24Y\right) = -495$$

First, we group the terms with X together. We group the terms with Y together and bring the constant on the other side of the equation.

$$\left(X^2 - 40X + \left(\frac{-40}{2}\right)^2\right) + \left(Y^2 + 24Y + \left(\frac{24}{2}\right)^2\right) = -495 + \left(\frac{-40}{2}\right)^2 + \left(\frac{24}{2}\right)^2$$

$$\left(X^2 - 40X + 400\right) + \left(Y^2 + 24Y + 144\right) = 49$$
$$(X - 20)(X - 20) + (Y + 12)(Y + 12) = 49$$
$$(X - 20)^2 + (Y + 12)^2 = 49$$

Once we complete the squares, the X's and Y's are perfect squares. We factor each.

$$(X - 20)^2 + (Y + 12)^2 = 49$$

This is our standard equation of circle.

» We also have standard equation of ellipse. When $a^2 > b^2$ and $\dfrac{X^2}{a^2}$, the ellipse is a horizontal ellipse. The horizontal ellipse is longer horizontally than vertically. The equation describes the horizontal ellipse.

$$\frac{(X - h)^2}{a^2} + \frac{(Y - k)^2}{b^2} = 1$$

Center: (h, k)
Foci: $(h - c, k)$ and $(h + c, k)$
$$c^2 = a^2 - b^2$$
Vertices: $(h - a, k)$ and $(h + a, k)$

We find c by applying $c^2 = a^2 - b^2$.

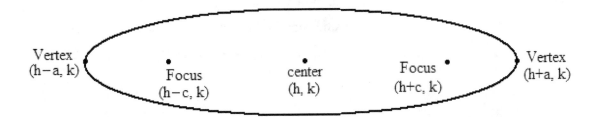

Vertex (h−a, k)   Focus (h−c, k)   center (h, k)   Focus (h+c, k)   Vertex (h+a, k)

» The ellipse is vertical when a^2 is underneath the Y^2.

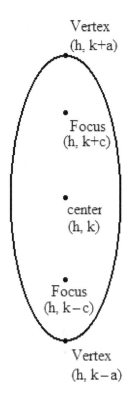

Vertex
(h, k+a)

• Focus
(h, k+c)

• center
(h, k)

• Focus
(h, k−c)

Vertex
(h, k−a)

$$\frac{(X-h)^2}{b^2} + \frac{(Y-k)^2}{a^2} = 1$$

where $a^2 > b^2$

Center: (h, k)
Foci: (h, k−c), (h, k+c)
Vertices: (h, k−a), (h, k+a)

Example 5:
Find the center, foci, and vertices
of the ellipse.

$$\frac{(X+21)^2}{25} + \frac{(Y-12)^2}{9} = 1$$

We have a horizontal ellipse.

$$\frac{(X-h)^2}{a^2} + \frac{(Y-k)^2}{b^2} = 1$$

We apply the ellipse formula.

$$\frac{(X-(-21))^2}{5^2} + \frac{(Y-12)^2}{3^2} = 1$$

We identify the values of h, k, and p.

Center: $(h,k) = (-21, 12)$

Foci: $(h-c, k)$ and $(h+c, k)$
$$c^2 = a^2 - b^2$$
$$c^2 = 25 - 9$$

For the focus, we find the value of
c by applying the formula
$$c^2 = a^2 - b^2$$

$$c^2 = 16$$
$$c = 4$$
$$(-25,12) \text{ and } (-17,12)$$

Vertices: $(h-a,k)$, $(h+a,k)$ For vertices, the value of a is 5.
$$(-26,12), (-16,12)$$

» We may not always get an ellipse equation in $\dfrac{(X-h)^2}{a^2} + \dfrac{(Y-k)^2}{b^2} = 1$ format. In this case, we will apply the completing the square method to convert to the standard equation of ellipse $\dfrac{(X-h)^2}{a^2} + \dfrac{(Y-k)^2}{b^2} = 1$.

Example 6:
Write the standard equation of ellipse
by completing the square.

$$2X^2 + 60X + 5Y^2 - 110Y + 1046 = 0$$

$$\left(2X^2 + 60X\right) + \left(5Y^2 - 110Y\right) = -1046$$ First, we group the X's together. We group the Y's together. We bring the constant to the other side.

$$2\left(X^2 + 30X\right) + 5\left(Y^2 - 22Y\right) = -1046$$ We factor out 2 from the group with X variables. We factor out 5 from the group with Y variables.

$$2\left(X^2 + 30X + \left(\frac{30}{2}\right)^2\right) + 5\left(Y^2 - 22Y + \left(\frac{-22}{2}\right)^2\right) = -1046 + 2\left(\frac{30}{2}\right)^2 + 5\left(\frac{-22}{2}\right)^2$$

$$2\left(X^2 + 30X + 225\right) + 5\left(Y^2 - 22Y + 121\right) = -1046 + 2(225) + 5(121)$$

$$2\left(X^2 + 30X + 225\right) + 5\left(Y^2 - 22Y + 121\right) = 9$$

$$2(X + 15)^2 + 5(Y - 11)^2 = 9$$

$$\frac{2(X+15)^2}{9} + \frac{5(Y-11)^2}{9} = 1$$

» Now, we will study the hyperbolas. We also have equations for the hyperbolas.

$$\frac{(X-h)^2}{a^2} - \frac{(Y-k)^2}{b^2} = 1$$ If X^2 comes first, it is a horizontal hyperbola.

Center: (h,k)

Foci: $(h-c,k)$ and $(h+c,k)$ We find the c by applying $b^2 = c^2 - a^2$.
$$b^2 = c^2 - a^2$$

Vertices: $(h-a,k)$ and $(h+a,k)$

We also have asymptotes. For horizontal hyperbola, asymptote formulas are
$$(Y-k) = \frac{b}{a}(X-h) \text{ and } (Y-k) = -\frac{b}{a}(X-h)$$

Horizontal hyperbolas

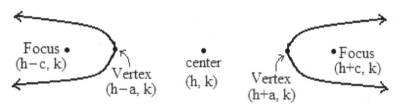

Asymptotes of horizontal hyperbola

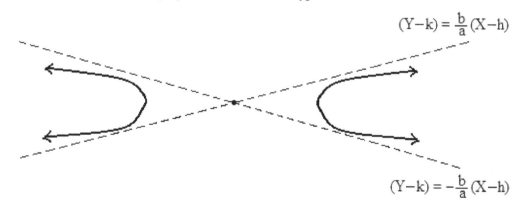

$$\frac{(Y-k)^2}{a^2} - \frac{(X-h)^2}{b^2} = 1$$ If Y^2 comes first, it is a vertical hyperbola.

Center: (h, k)

Foci: $(h, k - c)$ and $(h, k + c)$ We find c by applying $b^2 = c^2 - a^2$.
$$b^2 = c^2 - a^2$$

Vertices: $(h, k - a)$ and $(h, k + a)$

The asymptote formulas of vertical hyperbolas are $(Y - k) = \dfrac{a}{b}(X - h)$ and

$(Y - k) = -\dfrac{a}{b}(X - h)$.

$(Y - k) = \dfrac{a}{b}(X - h)$

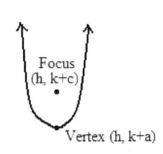

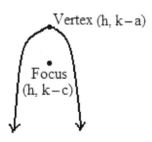

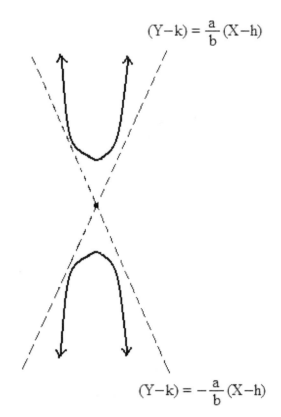

$(Y - k) = -\dfrac{a}{b}(X - h)$

Example 7:
Find the center, foci, and vertices
of the hyperbola.

$$\frac{(Y - 11)^2}{100} - \frac{(X + 8)^2}{4} = 1$$ We have a vertical hyperbola.

$$\frac{(Y - k)^2}{a^2} - \frac{(X - h)^2}{b^2} = 1$$ We apply the hyperbola formula.

$$\frac{(Y-11)^2}{10^2} - \frac{(X-(-8))^2}{2^2} = 1$$ We identify the values of h, k, a, and b.

Center: $(h,k) = (-8,11)$

Foci: $(h, k-c)$ and $(h, k+c)$ For foci, we apply $b^2 = c^2 - a^2$ to
$$b^2 = c^2 - a^2$$ find the value of c.
$$4 = c^2 - 100$$
$$c = \sqrt{104} = 2\sqrt{26}$$
$$(-8, 11 - 2\sqrt{26}), (-8, 11 + 2\sqrt{26})$$

Vertices: $(h, k-a)$ and $(h, k+a)$
$$(-8, 11-10) \text{ and } (-8, 11+10)$$
$$(-8,1) \text{ and } (-8,21)$$

Example 8:
Find the asymptotes of the hyperbola.

$$\frac{(X+2)^2}{36} - \frac{(Y-8)^2}{100} = 1$$ We have a horizontal hyperbola.

$$(Y-k) = \pm\frac{b}{a}(X-h)$$ We apply the asymptote formula for
a horizontal hyperbola.

$$(Y-8) = \pm\frac{10}{6}(X+2)$$

$$(Y-8) = \pm\frac{5}{3}(X+2)$$

$$(Y-8) = \frac{5}{3}(X+2) \text{ and } (Y-8) = -\frac{5}{3}(X+2)$$
$$Y = \frac{5}{3}X + \frac{34}{3} \text{ and } Y = -\frac{5}{3}X + \frac{14}{3}$$

» Hyperbola is not always in $\dfrac{(X-h)^2}{a^2} - \dfrac{(Y-k)^2}{b^2} = 1$ form. We apply the completing-the-

square method to express hyperbola in $\dfrac{(X-h)^2}{a^2} - \dfrac{(Y-k)^2}{b^2} = 1$ format.

Example 9:
Write the standard equation of hyperbola
by completing the square.

$$25X^2 - 400X - 4Y^2 - 8Y + 1{,}496 = 0$$

$$\left(25X^2 - 400X\right) + \left(-4Y^2 - 8Y\right) = -1{,}496$$

We group the terms with X together. We group Y's. We bring the constant to the other side of the equation.

$$25\left(X^2 - 16X\right) - 4\left(Y^2 + 2Y\right) = -1{,}496$$

We factor each group so that we have 1 as coefficient of X^2 and Y^2.

$$25\left(X^2 - 16X + \left(\frac{-16}{2}\right)^2\right) - 4\left(Y^2 + 2Y + \left(\frac{2}{2}\right)^2\right) = -1{,}496 + 25\left(\frac{-16}{2}\right)^2 - 4\left(\frac{2}{2}\right)^2$$

$$25\left(X^2 - 16X + 64\right) - 4\left(Y^2 + 2Y + 1\right) = 100$$

$$25(X - 8)(X - 8) - 4(Y + 1)(Y + 1) = 100$$

$$25(X - 8)^2 - 4(Y + 1)^2 = 100$$

$$\frac{25(X - 8)^2}{100} - \frac{4(Y + 1)^2}{100} = \frac{100}{100}$$

We set one side equal to 1.

$$\frac{(X - 8)^2}{4} - \frac{(Y + 1)^2}{25} = 1$$

» We classify the conic sections in equation format $aX^2 + cY^2 + bX + dY + e = 0$ (In this equation, both a and c cannot equal to 0 at the same time). We look for certain characteristics.

In the equation $aX^2 + cY^2 + bX + dY + e = 0$, we classify the equation a parabola if $ac = 0$.

In the equation $aX^2 + cY^2 + bX + dY + e = 0$, we classify the equation a circle if $a = c$.

In the equation $aX^2 + cY^2 + bX + dY + e = 0$, we classify the equation an ellipse if $ac > 0$ and $a \neq c$.

In the equation $aX^2 + cY^2 + bX + dY + e = 0$, we classify the equation a hyperbola if $ac < 0$.

Example 10:
Classify the conic section.

$$-5X^2 + 8Y^2 + 3X - 2Y + 3 = 0$$

$aX^2 + cY^2 + bX + dY + e = 0$ We identify the coefficients.
$a = -5$, $b = 3$, $c = 8$, $d = -2$, $e = 3$

$(-5)(8) < 0$
$ac < 0$

Answer: Hyperbola

Chapter 3   Conic sections

1. Find the vertex, focus, and directrix. $X^2 = -12Y$	2. Find the vertex, focus, and directrix. $Y^2 = 40X$
3. Find the vertex, focus, and directrix. $\left(X - \sqrt{2}\right)^2 = 3(Y - 1)$	4. Find the vertex, focus, and directrix. $\dfrac{\left(Y - \sqrt{5}\right)^2}{2} = (X - 8)$
5. Find the vertex, focus, and directrix. $\dfrac{(X + 7)^2}{11} = (Y - 1)$	6. Find the vertex, focus, and directrix. $\dfrac{(X - 1)^2}{-5} = (Y - 2)$
7. Find the vertex, focus, and directrix. $(X + 1)^2 = 2\left(Y - \sqrt{7}\right)$	8. Find the vertex, focus, and directrix. $\left(Y - \sqrt{2}\right)^2 = -5(X - 1)$
9. Find the vertex, focus, and directrix. $\left(X - \sqrt{3}\right)^2 = 4(Y - 1)$	10. Find the vertex, focus, and directrix. $\left(Y + \sqrt{2}\right)^2 = 11(X - 5)$

Chapter 3   Conic sections

11. Find the center and radius of the circle.	12. Find the center and radius of the circle.
$$\left(X+7\right)^2 + \left(Y-\sqrt{5}\right)^2 = 25$$	$$\left(X-\sqrt{2}\right)^2 + \left(Y+2\right)^2 = 13$$
13. Write the standard equation of circle by completing the square. $$X^2 - 4X + Y^2 + 24Y + 127 = 0$$	14. Write the standard equation of circle by completing the square. $$X^2 + 2\sqrt{5}X + Y^2 - 2Y - 2 = 0$$
15. Write the standard equation of circle by completing the square. $$X^2 + 10X + Y^2 - 2\sqrt{2}Y - 37 = 0$$	16. Write the standard equation of circle by completing the square. $$X^2 - 20X + Y^2 + 2Y + 64 = 0$$
17. Find the center, foci, and vertices of the ellipse. $$\frac{\left(X-2\right)^2}{25} + \frac{\left(Y+1\right)^2}{4} = 1$$	18. Find the center, foci, and vertices of the ellipse. $$\frac{\left(X+\sqrt{2}\right)^2}{2} + \frac{\left(Y-5\right)^2}{49} = 1$$
19. Write the standard equation of ellipse by completing the square. $$11X^2 + 44X + 4Y^2 - 8Y + 4 = 0$$	20. Write the standard equation of ellipse by completing the square. $$X^2 - 2X + 25Y^2 + 50\sqrt{5}Y + 101 = 0$$

Chapter 3   Conic sections

21. Write the standard equation of ellipse by completing the square. $$7X^2 + 14X + 3Y^2 - 6\sqrt{11}Y + 19 = 0$$	22. Write the standard equation of ellipse by completing the square. $$2X^2 - 4\sqrt{13}X + 5Y^2 - 10Y + 21 = 0$$
23. Find the center, foci, and vertices of the hyperbola. $$\frac{(X - \sqrt{2})^2}{3} - \frac{(Y - 5)^2}{5} = 1$$	24. Find the center, foci, and vertices of the hyperbola. $$\frac{(Y - \sqrt{3})^2}{2} - \frac{(X - \sqrt{2})^2}{7} = 1$$
25. Find the asymptotes of the hyperbola. $$\frac{(Y - 2)^2}{25} - (X + 1)^2 = 1$$	26. Find the asymptotes of the hyperbola. $$\frac{(X - 4)^2}{36} - \frac{(Y + 1)^2}{4} = 1$$
27. Write the standard equation of hyperbola by completing the square. $$2X^2 - 4X - 3Y^2 - 6\sqrt{3}Y - 13 = 0$$	28. Write the standard equation of hyperbola by completing the square. $$X^2 + 2\sqrt{2}X - 2Y^2 + 4\sqrt{5}Y - 12 = 0$$
29. Write the standard equation of hyperbola by completing the square. $$2Y^2 - 4\sqrt{2}Y - 7X^2 + 14X - 17 = 0$$	30. Write the standard equation of hyperbola by completing the square. $$5X^2 - 10\sqrt{11}X - 11Y^2 + 44Y - 44 = 0$$

Chapter 3   Conic sections

31. Classify the conic section. $$5X^2 + 3Y^2 + 3X + 2Y - 14 = 0$$	32. Classify the conic section. $$7X^2 + 7Y^2 - 2X - 4Y - 1 = 0$$
33. Classify the conic section. $$8X^2 + 7Y^2 + 4X - 4Y + 12 = 0$$	34. Classify the conic section. $$-2X^2 + 3Y^2 - 4X + Y - 1 = 0$$
35. Classify the conic section. $$X^2 - 2X + Y + 3 = 0$$	36. Classify the conic section. $$Y^2 + 2Y + X - 7 = 0$$
37. Classify the conic section. $$-2X^2 - 5Y^2 + 5X - 2Y + 1 = 0$$	38. Classify the conic section. $$5X^2 + 5Y^2 - 2X - Y - 5 = 0$$
39. Classify the conic section. $$8X^2 + 8Y^2 - 4X - 3Y + 11 = 0$$	40. Classify the conic section. $$-3X^2 + 3Y^2 + 2X - 2Y + 1 = 0$$

Chapter 3 Conic sections
Answer key

1. $(X-h)^2 = 4(p)(Y-k)$
   $(X-0)^2 = 4(-3)(Y-0)$
   Vertex : $(0,0)$
   Focus : $(0,-3)$
   Directrix : $Y = 3$

2. $(Y-k)^2 = 4(p)(X-h)$
   $(Y-0)^2 = 4(10)(X-0)$
   Vertex : $(0,0)$
   Focus : $(10,0)$
   Directrix : $X = -10$

3. $(X-h)^2 = 4(p)(Y-k)$
   $(X-\sqrt{2})^2 = 4\left(\dfrac{3}{4}\right)(Y-1)$
   Vertex : $(\sqrt{2},1)$
   Focus : $\left(\sqrt{2},\dfrac{7}{4}\right)$
   Directrix : $Y = \dfrac{1}{4}$

4. $(Y-k)^2 = 4(p)(X-h)$
   $(Y-\sqrt{5})^2 = 4\left(\dfrac{1}{2}\right)(X-8)$
   Vertex : $(8,\sqrt{5})$
   Focus : $\left(\dfrac{17}{2},\sqrt{5}\right)$
   Directrix : $X = \dfrac{15}{2}$

5. $(X-h)^2 = 4(p)(Y-k)$
   $(X+7)^2 = 4\left(\dfrac{11}{4}\right)(Y-1)$
   Vertex : $(-7,1)$
   Focus : $\left(-7,\dfrac{15}{4}\right)$
   Directrix : $Y = \dfrac{-7}{4}$

6. $(X-h)^2 = 4(p)(Y-k)$
   $(X-1)^2 = 4\left(\dfrac{-5}{4}\right)(Y-2)$
   Vertex : $(1,2)$
   Focus : $\left(1,\dfrac{3}{4}\right)$
   Directrix : $Y = \dfrac{13}{4}$

7. $(X-h)^2 = 4(p)(Y-k)$
   $(X+1)^2 = 4\left(\dfrac{1}{2}\right)(Y-\sqrt{7})$
   Vertex : $(-1,\sqrt{7})$
   Focus : $\left(-1,\sqrt{7}+\dfrac{1}{2}\right)$
   Directrix : $Y = \sqrt{7}-\dfrac{1}{2}$

8. $(Y-k)^2 = 4(p)(X-h)$
   $(Y-\sqrt{2})^2 = 4\left(\dfrac{-5}{4}\right)(X-1)$
   Vertex : $(1,\sqrt{2})$
   Focus : $\left(\dfrac{-1}{4},\sqrt{2}\right)$
   Directrix : $X = \dfrac{9}{4}$

9. $(X-h)^2 = 4(p)(Y-k)$

$(X-\sqrt{3})^2 = 4(1)(Y-1)$

$Vertex: (\sqrt{3},1)$

$Focus: (\sqrt{3},2)$

$Directrix: Y = 0$

10. $(Y-k)^2 = 4(p)(X-h)$

$(Y+\sqrt{2})^2 = 4\left(\frac{11}{4}\right)(X-5)$

$Vertex: (5,-\sqrt{2})$

$Focus: \left(\frac{31}{4},-\sqrt{2}\right)$

$Directrix: X = \frac{9}{4}$

11. $(X-h)^2 + (Y-k)^2 = r^2$

$(X-(-7))^2 + (Y-\sqrt{5})^2 = 5^2$

$Center: (-7,\sqrt{5})$ and $radius = 5$

12. $(X-h)^2 + (Y-k)^2 = r^2$

$(X-\sqrt{2})^2 + (Y-(-2))^2 = (\sqrt{13})^2$

$Center: (\sqrt{2},-2)$ and $radius = \sqrt{13}$

13. $X^2 - 4X + Y^2 + 24Y + 127 = 0$

$(X^2 - 4X) + (Y^2 + 24Y) = -127$

$\left(X^2 - 4X + \left(\frac{-4}{2}\right)^2\right) + \left(Y^2 + 24Y + \left(\frac{24}{2}\right)^2\right) = -127 + 4 + 144$

$(X^2 - 4X + 4) + (Y^2 + 24Y + 144) = 21$

$(X-2)^2 + (Y+12)^2 = 21$

14. $X^2 + 2\sqrt{5}X + Y^2 - 2Y - 2 = 0$

$(X^2 + 2\sqrt{5}X) + (Y^2 - 2Y) = 2$

$\left(X^2 + 2\sqrt{5}X + \left(\frac{2\sqrt{5}}{2}\right)^2\right) + \left(Y^2 - 2Y + \left(\frac{-2}{2}\right)^2\right) = 2 + 5 + 1$

$(X^2 + 2\sqrt{5}X + 5) + (Y^2 - 2Y + 1) = 8$

$(X+\sqrt{5})^2 + (Y-1)^2 = 8$

15. $X^2 + 10X + Y^2 - 2\sqrt{2}Y - 37 = 0$

$(X^2 + 10X) + (Y^2 - 2\sqrt{2}Y) = 37$

$\left(X^2 + 10X + \left(\frac{10}{2}\right)^2\right) + \left(Y^2 - 2\sqrt{2}Y + \left(\frac{-2\sqrt{2}}{2}\right)^2\right) = 37 + 25 + 2$

$(X^2 + 10X + 25) + (Y^2 - 2\sqrt{2}Y + 2) = 64$

$(X+5)^2 + (Y-\sqrt{2})^2 = 64$

16. $X^2 - 20X + Y^2 + 2Y + 64 = 0$

$(X^2 - 20X) + (Y^2 + 2Y) = -64$

$\left(X^2 - 20X + \left(\frac{-20}{2}\right)^2\right) + \left(Y^2 + 2Y + \left(\frac{2}{2}\right)^2\right) = -64 + 100 + 1$

$(X^2 - 20X + 100) + (Y^2 + 2Y + 1) = 37$

$(X-10)^2 + (Y+1)^2 = 37$

17. $\frac{(X-2)^2}{5^2} + \frac{(Y+1)^2}{2^2} = 1$

$Center: (2,-1)$

$Foci: c^2 = a^2 - b^2 \rightarrow c = \sqrt{21}$

$(2-\sqrt{21},-1)(2+\sqrt{21},-1)$

$Vertices: (-3,-1)(7,-1)$

18. $\frac{(X+\sqrt{2})^2}{(\sqrt{2})^2} + \frac{(Y-5)^2}{7^2} = 1$

$Center: (-\sqrt{2},5)$

$Foci: c^2 = a^2 - b^2 \rightarrow c = \sqrt{47}$

$(-\sqrt{2},5-\sqrt{47})(-\sqrt{2},5+\sqrt{47})$

$Vertices: (-\sqrt{2},-2)(-\sqrt{2},12)$

19. $11X^2 + 44X + 4Y^2 - 8Y + 4 = 0$

$11(X^2 + 4X) + 4(Y^2 - 2Y) = -4$

$11\left(X^2 + 4X + \left(\frac{4}{2}\right)^2\right) + 4\left(Y^2 - 2Y + \left(\frac{-2}{2}\right)^2\right) = -4 + 44 + 4$

$11(X^2 + 4X + 4) + 4(Y^2 - 2Y + 1) = 44$

$11(X+2)^2 + 4(Y-1)^2 = 44$

$\frac{(X+2)^2}{4} + \frac{(Y-1)^2}{11} = 1$

20. $X^2 - 2X + 25Y^2 + 50\sqrt{5}Y + 101 = 0$

$(X^2 - 2X) + 25(Y^2 + 2\sqrt{5}Y) = -101$

$\left(X^2 - 2X + \left(\frac{-2}{2}\right)^2\right) + 25\left(Y^2 + 2\sqrt{5}Y + \left(\frac{2\sqrt{5}}{2}\right)^2\right) = -101 + 1 + 125$

$(X^2 - 2X + 1) + 25(Y^2 + 2\sqrt{5}Y + 5) = 25$

$(X - 1)^2 + 25(Y + \sqrt{5})^2 = 25$

$\dfrac{(X-1)^2}{25} + (Y + \sqrt{5})^2 = 1$

21. $7X^2 + 14X + 3Y^2 - 6\sqrt{11}Y + 19 = 0$

$7(X^2 + 2X) + 3(Y^2 - 2\sqrt{11}Y) = -19$

$7\left(X^2 + 2X + \left(\frac{2}{2}\right)^2\right) + 3\left(Y^2 - 2\sqrt{11}Y + \left(\frac{-2\sqrt{11}}{2}\right)^2\right) = -19 + 7 + 33$

$7(X^2 + 2X + 1) + 3(Y^2 - 2\sqrt{11}Y + 11) = 21$

$\dfrac{(X+1)^2}{3} + \dfrac{(Y - \sqrt{11})^2}{7} = 1$

22. $2X^2 - 4\sqrt{13}X + 5Y^2 - 10Y + 21 = 0$

$2(X^2 - 2\sqrt{13}X) + 5(Y^2 - 2Y) = -21$

$2\left(X^2 - 2\sqrt{13}X + \left(\frac{-2\sqrt{13}}{2}\right)^2\right) + 5\left(Y^2 - 2Y + \left(\frac{-2}{2}\right)^2\right) = -21 + 26 + 5$

$2(X^2 - 2\sqrt{13}X + 13) + 5(Y^2 - 2Y + 1) = 10$

$2(X - \sqrt{13})^2 + 5(Y - 1)^2 = 10$

$\dfrac{(X - \sqrt{13})^2}{5} + \dfrac{(Y - 1)^2}{2} = 1$

23. $\dfrac{(X - \sqrt{2})^2}{(\sqrt{3})^2} - \dfrac{(Y - 5)^2}{(\sqrt{5})^2} = 1$

Center : $(\sqrt{2}, 5)$

Foci : $b^2 = c^2 - a^2 \to c = 2\sqrt{2}$

$(-\sqrt{2}, 5)(3\sqrt{2}, 5)$

Vertices : $(\sqrt{2} - \sqrt{3}, 5)(\sqrt{2} + \sqrt{3}, 5)$

24. $\dfrac{(Y - \sqrt{3})^2}{(\sqrt{2})^2} - \dfrac{(X - \sqrt{2})^2}{(\sqrt{7})^2} = 1$

Center : $(\sqrt{2}, \sqrt{3})$

Foci : $b^2 = c^2 - a^2 \to c = 3$

$(\sqrt{2}, \sqrt{3} - 3)(\sqrt{2}, \sqrt{3} + 3)$

Vertices : $(\sqrt{2}, \sqrt{3} - \sqrt{2})(\sqrt{2}, \sqrt{3} + \sqrt{2})$

25. $\dfrac{(Y - 2)^2}{5^2} - \dfrac{(X + 1)^2}{1^2} = 1$

$(Y - 2) = \pm\dfrac{5}{1}(X + 1)$

$(Y - 2) = 5(X + 1) \qquad Y - 2 = -5(X + 1)$

$Y - 2 = 5X + 5 \quad \text{and} \quad Y - 2 = -5X - 5$

$Y = 5X + 7 \qquad\qquad Y = -5X - 3$

26. $\dfrac{(X - 4)^2}{6^2} - \dfrac{(Y + 1)^2}{2^2} = 1$

$(Y + 1) = \pm\dfrac{1}{3}(X - 4)$

$(Y + 1) = \dfrac{1}{3}(X - 4) \quad (Y + 1) = \dfrac{-1}{3}(X - 4)$

$Y + 1 = \dfrac{1}{3}X - \dfrac{4}{3} \qquad Y + 1 = \dfrac{-1}{3}X + \dfrac{4}{3}$

$Y = \dfrac{1}{3}X - \dfrac{7}{3} \qquad Y = \dfrac{-1}{3}X + \dfrac{1}{3}$

27. $2X^2 - 4X - 3Y^2 - 6\sqrt{3}Y - 13 = 0$

$2(X^2 - 2X) - 3(Y^2 + 2\sqrt{3}Y) = 13$

$2\left(X^2 - 2X + \left(\frac{-2}{2}\right)^2\right) - 3\left(Y^2 + 2\sqrt{3}Y + \left(\frac{2\sqrt{3}}{2}\right)^2\right) = 13 + 2 - 9$

$2(X^2 - 2X + 1) - 3(Y^2 + 2\sqrt{3}Y + 3) = 6$

$2(X - 1)^2 - 3(Y + \sqrt{3})^2 = 6$

$\dfrac{(X - 1)^2}{3} - \dfrac{(Y + \sqrt{3})^2}{2} = 1$

28. $X^2 + 2\sqrt{2}X - 2Y^2 + 4\sqrt{5}Y - 12 = 0$

$\left(X^2 + 2\sqrt{2}X\right) - 2\left(Y^2 - 2\sqrt{5}Y\right) = 12$

$\left(X^2 + 2\sqrt{2}X + \left(\frac{2\sqrt{2}}{2}\right)^2\right) - 2\left(Y^2 - 2\sqrt{5}Y + \left(\frac{-2\sqrt{5}}{2}\right)^2\right) = 12 + 2 - 10$

$\left(X^2 + 2\sqrt{2}X + 2\right) - 2\left(Y^2 - 2\sqrt{5}Y + 5\right) = 4$

$\left(X + \sqrt{2}\right)^2 - 2\left(Y - \sqrt{5}\right)^2 = 4$

$\dfrac{\left(X + \sqrt{2}\right)^2}{4} - \dfrac{\left(Y - \sqrt{5}\right)^2}{2} = 1$

29. $2Y^2 - 4\sqrt{2}Y - 7X^2 + 14X - 17 = 0$

$2\left(Y^2 - 2\sqrt{2}Y\right) - 7\left(X^2 - 2X\right) = 17$

$2\left(Y^2 - 2\sqrt{2}Y + \left(\frac{-2\sqrt{2}}{2}\right)^2\right) - 7\left(X^2 - 2X + \left(\frac{-2}{2}\right)^2\right) = 17 + 4 - 7$

$2\left(Y^2 - 2\sqrt{2}Y + 2\right) - 7\left(X^2 - 2X + 1\right) = 14$

$\dfrac{\left(Y - \sqrt{2}\right)^2}{7} - \dfrac{(X - 1)^2}{2} = 1$

30.

$5X^2 - 10\sqrt{11}X - 11Y^2 + 44Y - 44 = 0$

$5\left(X^2 - 2\sqrt{11}X\right) - 11\left(Y^2 - 4Y\right) = 44$

$5\left(X^2 - 2\sqrt{11}X + \left(\frac{-2\sqrt{11}}{2}\right)^2\right) - 11\left(Y^2 - 4Y + \left(\frac{-4}{2}\right)^2\right) = 44 + 55 - 44$

$5\left(X^2 - 2\sqrt{11}X + 11\right) - 11\left(Y^2 - 4Y + 4\right) = 55$

$5\left(X - \sqrt{11}\right)^2 - 11(Y - 2)^2 = 55$

$\dfrac{\left(X - \sqrt{11}\right)^2}{11} - \dfrac{(Y - 2)^2}{5} = 1$

31. $aX^2 + cY^2 + bX + dY + e = 0$

$ac > 0$ and $a \neq c \rightarrow$ ellipse

$5X^2 + 3Y^2 + 3X + 2Y - 14 = 0$

$(5)(3) > 0$ and $5 \neq 3 \rightarrow$ Ellipse

32. $aX^2 + cY^2 + bX + dY + e = 0$

$a = c \rightarrow$ circle

$7X^2 + 7Y^2 - 2X - 4Y - 1 = 0$

$7 = 7$

Circle

33. $aX^2 + cY^2 + bX + dY + e = 0$

$ac > 0$ and $a \neq c \rightarrow$ ellipse

$8X^2 + 7Y^2 + 4X - 4Y + 12 = 0$

$(8)(7) > 0$ and $8 \neq 7 \rightarrow$ Ellipse

34. $aX^2 + cY^2 + bX + dY + e = 0$

$ac < 0 \rightarrow$ hyperbola

$-2X^2 + 3Y^2 - 4X + Y - 1 = 0$

$(-2)(3) < 0 \rightarrow$ Hyperbola

35. $aX^2 + cY^2 + bX + dY + e = 0$

$ac = 0 \rightarrow$ parabola

$X^2 + (0)Y^2 - 2X + Y + 3 = 0$

$(1)(0) = 0 \rightarrow$ Parabola

36. $aX^2 + cY^2 + bX + dY + e = 0$

$ac = 0 \rightarrow$ parabola

$(0)X^2 + Y^2 + X + 2Y - 7 = 0$

$(0)(1) = 0 \rightarrow$ Parabola

37. $aX^2 + cY^2 + bX + dY + e = 0$

$ac > 0$ and $a \neq c \rightarrow$ ellipse

$-2X^2 - 5Y^2 + 5X - 2Y + 1 = 0$

$(-2)(-5) > 0$ and $-2 \neq -5$

Ellipse

38. $aX^2 + cY^2 + bX + dY + e = 0$

$a = c \rightarrow$ circle

$5X^2 + 5Y^2 - 2X - Y - 5 = 0$

$5 = 5 \rightarrow$ Circle

39. $aX^2 + cY^2 + bX + dY + e = 0$

$a = c \rightarrow$ circle

$8X^2 + 8Y^2 - 4X - 3Y + 11 = 0$

$8 = 8 \rightarrow$ Circle

40. $aX^2 + cY^2 + bX + dY + e = 0$

$ac < 0 \rightarrow$ hyperbola

$-3X^2 + 3Y^2 + 2X - 2Y + 1 = 0$

$(-3)(3) < 0 \rightarrow$ Hyperbola

Chapter 4   Finding zeros

▶ Zeros are the points where the graph crosses the x-axis. Zeros can be referred to as solutions, roots, or x-intercepts. To find zero, we can substitute 0 in Y, since the zeros all have 0 for the y-coordinate.

In a quadratic equation, we can find the zeros (roots or solutions) by applying factoring method. We factor the quadratic equation and set each factor equal to zero. Then, we solve for X, which will give us the zeros.

Example 1:
Find the zeros.

$$X^2 + 8X + 15 = 0$$ We factor the quadratic equation.

The factors of 15: $1 \cdot 15$ We find the factors of the constant. Because
$3 \cdot 5$ we have plus sign in front of the constant, we add the factors to see which one has sum same as the middle coefficient.

$$(X + 3)(X + 5) = 0$$ We have our factors.

$$X + 3 = 0 \quad \text{and} \quad X + 5 = 0$$ We set each factor equal to 0 because, if either factor is equal to 0, the entire equation equals zero.

$$X = -3 \quad \text{and} \quad -5$$ We have our zeros.

» The quadratic equation is not always factorable. In this case, we can apply the quadratic formula to find the value or values of the zeros.

$$aX^2 + bX + c = 0$$

Quadratic formula: $X = \dfrac{-b \pm \sqrt{b^2 - 4ac}}{2a}$

Example 2:
Apply the quadratic formula to find the zeros.

$$2X^2 - 5X + 1 = 0$$

$$aX^2 + bX + c = 0$$ We first identify the values of *a, b,* and *c.*
$$a = 2, \; b = -5, \; c = 1$$

$$X = \frac{-b \pm \sqrt{b^2 - 4ac}}{2a}$$

$$= \frac{-(-5) \pm \sqrt{(-5)^2 - 4(2)(1)}}{2(2)}$$ We substitute the corresponding values.

$$= \frac{5 \pm \sqrt{25 - 8}}{4} = \frac{5 \pm \sqrt{17}}{4}$$

$$X = \frac{5 - \sqrt{17}}{4} \quad \text{and} \quad \frac{5 + \sqrt{17}}{4}$$

» We may see a higher-degree polynomials. In this case, we apply the synthetic division to find the values of zeros.

Example 3:
Applying synthetic division, find the zeros if one of its zeros is 1.

$$X^3 - 4X^2 - 7X + 10 = 0$$

$$X^3 - 4X^2 - 7X + 10$$

$$\begin{array}{c|cccc} & 1 & -4 & -7 & 10 \\ \hline \end{array}$$

We write the coefficients of the equation.

We write the zero here. $\longrightarrow$

$$\begin{array}{c|cccc} 1 & 1 & -4 & -7 & 10 \\ \hline \end{array}$$

Because one of the zeros is 1, we write 1 as shown left.

$$
\begin{array}{r|rrrr}
1 & 1 & -4 & -7 & 10 \\
  &   & 1 &   &   \\
\hline
  & 1 &   &   &   \\
\end{array}
$$

We bring down the first coefficient and multiply by the 1. We put the product underneath the second column.

$$
\begin{array}{r|rrrr}
1 & 1 & -4 & -7 & 10 \\
  &   & 1 & -3 &   \\
\hline
  & 1 & -3 &   &   \\
\end{array}
$$

We combine the second column, which results in -3. We multiply the result by 1 and write the product underneath the third column.

$$
\begin{array}{r|rrrr}
1 & 1 & -4 & -7 & 10 \\
  &   & 1 & -3 & -10 \\
\hline
  & 1 & -3 & -10 &   \\
\end{array}
$$

We combine the third column, which results in -10. Then, we multiply the result by 1 and write the product underneath the fourth column.

$$
\begin{array}{r|rrrr}
1 & 1 & -4 & -7 & 10 \\
  &   & 1 & -3 & -10 \\
\hline
  & 1 & -3 & -10 & 0 \\
\end{array}
$$

We combine the fourth column. If the last column (the remainder) results in 0, the divisor is the zero. Therefore, 1 is the zero.

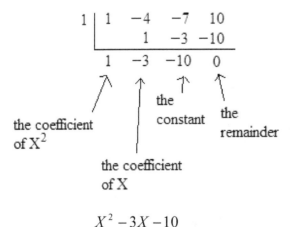

$$X^2 - 3X - 10$$

We can apply the factoring method to find the remaining zeros.

$$(X-5)(X+2) = 0$$
$$X = 5 \text{ and } -2$$

Zeros: 1, 5, − 2

» The intermediate value theorem verifies whether zero exists in the given interval. If the graph is continuous and the y-coordinates of two points are opposite signs, then the graph must cross the x-axis some place in between the two intervals. Therefore, we know that at least one zero exists in the interval.

We have two points $(2,-7)$ and $(5,2)$ of a continuous graph.

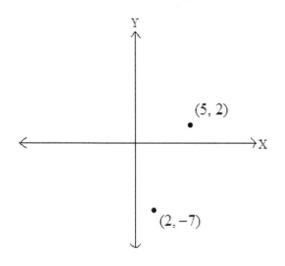

We have two points of a continuous function. Because the graph is continuous function, we have to cross the x-axis someplace between the interval (2, 5). We find that we cannot connect the two points without crossing someplace between the interval.

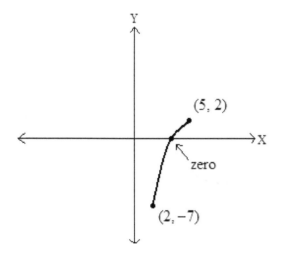

Based on the intermediate value theorem, the continuous graph must cross the x-axis (has a zero) between the two points if the y-coordinates are opposite signs.

Example 4:
Determine whether the graph crosses x-axis on the interval $[-2,3]$.

$$f(X) = 2X^3 + 5X^2 - 2X + 1$$
Interval: $[-2,3]$

We first determine whether the graph is continuous function. The graph is $f(X)$ is continuous on the interval $[-2,3]$. We apply the intermediate value theorem.

$f(-2) = 2(-2)^3 + 5(-2)^2 - 2(-2) + 1$ We substitute each endpoint in the

$f(3) = 2(3)^3 + 5(3)^2 - 2(3) + 1$ function to determine y-coordinate.

$f(-2) = 9$ The y-coordinates (9 and 94) are the

$f(3) = 94$ same signs.

Whether the graph crosses the x-axis on the interval $[-2,3]$ cannot be determined.

Example 5:
Determine whether the graph crosses x-axis on the interval $[-3,0]$.

$$f(X) = 2X^3 - 8X^2 - 14X + 20$$
Interval $[-3,0]$

We find that the graph is a continuous function. We apply the intermediate value theorem.

$f(-3) = 2(-3)^3 - 8(-3)^2 - 14(-3) + 20$ We substitute the endpoint in the

$f(0) = 2(0)^3 - 8(0)^2 - 14(0) + 20$ function to identify the y-coordinate.

$f(-3) = -64$ The y-coordinates $(-64$ and $20)$ are

$f(0) = 20$ the opposite signs.

By intermediate value theorem, the graph crosses the x-axis on the interval $[-3,0]$.

» Applying Descartes Rule of Signs, we can find the numbers of possible positive and negative zeros. For the number of positive zeros, we look for sign changes in the original equation. We start from the left and count the number of times the sign changes from positive to negative or negative to positive.

$$f(X) = \underbrace{X^3 - 4X^2}_{\substack{\text{sign} \\ \text{change}}} \underbrace{- 7X + 10}_{\substack{\text{sign} \\ \text{change}}}$$

Two sign changes indicate two positive zero. We also subtract the number of sign changes by 2's. So, we may have two or zero positive zeros (we cannot have negative number of positive zeros so we stop at 0).

To find the number of negative zeros, we substitute $-X$ into the original equation and find the number of sign changes. The number of sign changes indicates the number of possible negative zeros. We also subtract by 2's until we get 0 or just before the negative.

$$f(-X) = (-X)^3 - 4(-X)^2 - 7(-X) + 10 \qquad \text{We substitute } -X.$$

$$= \underbrace{-X^3 - 4X^2 + 7X}_{\substack{\text{sign} \\ \text{change}}} + 10$$

One sign change indicates one negative zero. Because we cannot subtract by 2's from this point, we have one negative zero in this equation.

Example 6:
Applying Descartes Rule of Signs, find the numbers of possible positive and negative zeros.

$$f(X) = X^5 - 6X^4 + 8X^3 - 20X^2 + 15X + 50$$

$$f(X) = \underbrace{X^5 - 6X^4}_{\substack{\text{sign} \\ \text{change}}} \underbrace{+ 8X^3}_{\substack{\text{sign} \\ \text{change}}} \underbrace{- 20X^2}_{\substack{\text{sign} \\ \text{change}}} \underbrace{+ 15X}_{\substack{\text{sign} \\ \text{change}}} + 50$$

We have four sign changes. We may have 4, 2, or 0 possible positive zeros.

4, 2, or 0 positive zeros

$$f(-X) = (-X)^5 - 6(-X)^4 + 8(-X)^3 - 20(-X)^2 + 15(-X) + 50 \qquad \text{We substitute } -X.$$

$$= -X^5 - 6X^4 - 8X^3 - 20X^2 \underbrace{- 15X + 50}_{\substack{\text{sign} \\ \text{change}}}$$

One sign change indicates one negative zero.

1 negative zero

» To find possible zeros, we divide the factors of the constant by the factors of the leading coefficient. We will obtain the potential zeros and we apply the synthetic division to see which one works.

Example 7:
Find the zeros.

$$X^4 - 5X^3 - 15X^2 + 5X + 14 = 0$$ The constant is 14 and the leading coefficient is 1.

The factors of the constant: $\pm 1, \pm 2, \pm 7, \pm 14$
The factors of the leading coefficient: ± 1

Possible zeros $= \dfrac{\pm 1, \pm 2, \pm 7, \pm 14}{\pm 1}$ We divide the factors of constant

by the factors of leading coefficient.

Possible zeros: $\pm 1, \pm 2, \pm 7, \pm 14$ We pick a number from possible zeros and apply the synthetic division. When we get 0 as a remainder, we have found the zero.

$$
\begin{array}{r|rrrrr}
1 & 1 & -5 & -15 & 5 & 14 \\
  &   & 1 & -4 & -19 & -14 \\
\hline
  & 1 & -4 & -19 & -14 & 0
\end{array}
$$
We verify that 1 is one of the zeros.

$$(X - 1)(X^3 - 4X^2 - 19X - 14)$$ We repeat the same process for the equation $X^3 - 4X^2 - 19X - 14$.

$$
\begin{array}{r|rrrr}
-1 & 1 & -4 & -19 & -14 \\
   &   & -1 & 5 & 14 \\
\hline
   & 1 & -5 & -14 & 0
\end{array}
$$
We verify that -1 is one of the zeros.

$$(X - 1)(X + 1)(X^2 - 5X - 14)$$

$$X^2 - 5X - 14 = (X - 7)(X + 2)$$ We can apply the factoring method.

$$(X - 1)(X + 1)(X - 7)(X + 2) = 0$$

Zeros: $1, -1, 7, -2$

Chapter 4   Finding zeros

1. Find the zeros. $$X^2 + 17X + 60 = 0$$	2. Find the zeros. $$X^2 + 9X - 52 = 0$$
3. Find the zeros. $$X^2 + 17X + 70 = 0$$	4. Find the zeros. $$X^2 - X - 132 = 0$$
5. Find the zeros. $$2X^2 + 9X - 35 = 0$$	6. Apply the quadratic formula to find the zeros. $$-5X^2 - 4X + 10 = 0$$
7. Apply the quadratic formula to find the zeros. $$2X^2 - 3X - 4 = 0$$	8. Apply the quadratic formula to find the zeros. $$4X^2 + 2X - 7 = 0$$
9. Apply the quadratic formula to find the zeros. $$3X^2 + 15X + 1 = 0$$	10. Apply the quadratic formula to find the zeros. $$-7X^2 + 2X + 1 = 0$$

Chapter 4   Finding zeros

11. Applying synthetic division, find the zeros if one of its zeros is −2. $X^3 + 4X^2 + X - 6 = 0$	12. Applying synthetic division, find the zeros if one of its zeros is 5. $X^3 + 3X^2 - 33X - 35 = 0$
13. Applying synthetic division, find the zeros if one of its zeros is 10. $X^3 + 6X^2 - 105X - 550 = 0$	14. Applying synthetic division, find the zeros if one of its zeros is −7. $X^3 + 4X^2 - 25X - 28 = 0$
15. Applying synthetic division, find the zeros if one of its zeros is 4. $X^3 + 12X^2 - 36X - 112 = 0$	16. Determine whether the graph crosses x-axis on the interval $\left[-5, -1\right]$. $f(X) = X^3 - 3X^2 - 10X + 24$ *Interval*: $\left[-5, -1\right]$
17. Determine whether the graph crosses x-axis on the interval $\left[1, 7\right]$. $f(X) = X^3 - 2X^2 - 40X - 64$ *Interval*: $\left[1, 7\right]$	18. Determine whether the graph crosses x-axis on the interval $\left[-15, -5\right]$. $f(X) = X^3 + 4X^2 - 52X + 80$ *Interval*: $\left[-15, -5\right]$
19. Determine whether the graph crosses x-axis on the interval $\left[0, 8\right]$. $f(X) = X^3 + 3X^2 - 18X - 40$ *Interval*: $\left[0, 8\right]$	20. Applying Descartes Rule of Signs, find the numbers of possible positive and negative zeros. $f(X) = X^4 + 8X^3 - 3X^2 - 130X - 200$

Chapter 4   Finding zeros

21. Applying Descartes Rule of Signs, find the numbers of possible positive and negative zeros. $f(X)= X^7 - 3X^5 - 4X^4 + 2X + 7$	22. Applying Descartes Rule of Signs, find the numbers of possible positive and negative zeros. $f(X)= -7X^5 + 7X^3 - 7X^2 + 5X + 1$
23. Applying Descartes Rule of Signs, find the numbers of possible positive and negative zeros. $f(X)= 3X^4 - 3X^3 + 2X^2 + 10X + 4$	24. Find the zeros. $X^3 + 13X^2 + 7X - 165 = 0$
25. Find the zeros. $X^3 + 4X^2 - 4X - 16 = 0$	26. Find the zeros. $X^3 - 6X^2 - X + 30 = 0$
27. Find the zeros. $2X^3 + X^2 - 26X - 40 = 0$	28. Find the zeros. $X^3 + 16X^2 - 73X - 952 = 0$
29. Find the zeros. $X^4 + 5X^3 - 30X^2 - 40X + 64 = 0$	30. Find the zeros. $X^4 - 4X^3 - 22X^2 + 4X + 21 = 0$

Chapter 4 Finding zeros
Answer key

1. $X^2 + 17X + 60 = 0$
   $(X+5)(X+12) = 0$
   $X + 5 = 0$ and $X + 12 = 0$
   $X = -5$ and $X = -12$

2. $X^2 + 9X - 52 = 0$
   $(X-4)(X+13) = 0$
   $X - 4 = 0$ and $X + 13 = 0$
   $X = 4$ and $X = -13$

3. $X^2 + 17X + 70 = 0$
   $(X+7)(X+10) = 0$
   $X + 7 = 0$ and $X + 10 = 0$
   $X = -7$ and $X = -10$

4. $X^2 - X - 132 = 0$
   $(X+11)(X-12) = 0$
   $X + 11 = 0$ and $X - 12 = 0$
   $X = -11$ and $X = 12$

5. $2X^2 + 9X - 35 = 0$
   $(2X-5)(X+7) = 0$
   $2X - 5 = 0$ and $X + 7 = 0$
   $X = \dfrac{5}{2}$ and $X = -7$

6. $-5X^2 - 4X + 10 = 0$
   $X = \dfrac{-(-4) \pm \sqrt{(-4)^2 - 4(-5)(10)}}{2(-5)}$
   $X = \dfrac{4 \pm 6\sqrt{6}}{-10}$
   $X = \dfrac{-2 + 3\sqrt{6}}{5}$ and $X = \dfrac{-2 - 3\sqrt{6}}{5}$

7. $2X^2 - 3X - 4 = 0$
   $X = \dfrac{-(-3) \pm \sqrt{(-3)^2 - 4(2)(-4)}}{2(2)}$
   $X = \dfrac{3 \pm \sqrt{41}}{4}$
   $X = \dfrac{3 - \sqrt{41}}{4}$ and $X = \dfrac{3 + \sqrt{41}}{4}$

8. $4X^2 + 2X - 7 = 0$
   $X = \dfrac{-2 \pm \sqrt{(2)^2 - 4(4)(-7)}}{2(4)}$
   $X = \dfrac{-2 \pm 2\sqrt{29}}{8}$
   $X = \dfrac{-1 - \sqrt{29}}{4}$ and $X = \dfrac{-1 + \sqrt{29}}{4}$

9. $3X^2 + 15X + 1 = 0$
   $X = \dfrac{-15 \pm \sqrt{(15)^2 - 4(3)(1)}}{2(3)}$
   $X = \dfrac{-15 \pm \sqrt{213}}{6}$
   $X = \dfrac{-15 - \sqrt{213}}{6}$ and $X = \dfrac{-15 + \sqrt{213}}{6}$

10. $-7X^2 + 2X + 1 = 0$
    $X = \dfrac{-2 \pm \sqrt{2^2 - 4(-7)(1)}}{2(-7)}$
    $X = \dfrac{-2 \pm 4\sqrt{2}}{-14}$
    $X = \dfrac{1 - 2\sqrt{2}}{7}$ and $X = \dfrac{1 + 2\sqrt{2}}{7}$

11.

$$
\begin{array}{r|rrrr}
-2 & 1 & 4 & 1 & -6 \\
& & -2 & -4 & 6 \\
\hline
& 1 & 2 & -3 & 0
\end{array}
$$

$$(X+2)(X^2+2X-3)=0$$
$$(X+2)(X-1)(X+3)=0$$
$$X=-2,\ X=1,\text{ and }X=-3$$

12.

$$
\begin{array}{r|rrrr}
5 & 1 & 3 & -33 & -35 \\
& & 5 & 40 & 35 \\
\hline
& 1 & 8 & 7 & 0
\end{array}
$$

$$(X-5)(X^2+8X+7)=0$$
$$(X-5)(X+7)(X+1)=0$$
$$X=5,\ X=-7,\text{ and }X=-1$$

13.

$$
\begin{array}{r|rrrr}
10 & 1 & 6 & -105 & -550 \\
& & 10 & 160 & 550 \\
\hline
& 1 & 16 & 55 & 0
\end{array}
$$

$$(X-10)(X^2+16X+55)$$
$$(X-10)(X+5)(X+11)$$
$$X=10,\ X=-5,\text{ and }X=-11$$

14.

$$
\begin{array}{r|rrrr}
-7 & 1 & 4 & -25 & -28 \\
& & -7 & 21 & 28 \\
\hline
& 1 & -3 & -4 & 0
\end{array}
$$

$$(X+7)(X^2-3X-4)=0$$
$$(X+7)(X+1)(X-4)=0$$
$$X=-7,\ X=-1,\text{ and }X=4$$

15.

$$
\begin{array}{r|rrrr}
4 & 1 & 12 & -36 & -112 \\
& & 4 & 64 & 112 \\
\hline
& 1 & 16 & 28 & 0
\end{array}
$$

$$(X-4)(X^2+16X+28)=0$$
$$(X-4)(X+2)(X+14)=0$$
$$X=4,\ X=-2,\text{ and }X=-14$$

16. $f(X)$ is continuous on the interval $[-5,-1]$.

$$f(-5)=-126$$
$$f(-1)=30$$

By intermediate value theorem, the graph crosses the x-axis on the interval $[-5,-1]$.

17. $f(X)$ is continuous on $[1,7]$.
$$f(1)=-105$$
$$f(7)=-99$$

Whether the graph crosses x-axis on the interval $[1,7]$ cannot be determined.

18. $f(X)$ is continuous on $[-15,-5]$.
$$f(-15)=-1,615$$
$$f(-5)=315$$

By intermediate value theorem, the graph crosses the x-axis on the interval $[-15,-5]$.

19. $f(X)$ is continuous on $[0,8]$.
$$f(0)=-40$$
$$f(8)=520$$

By intermediate value theorem, the graph crosses the x-axis on the interval $[0,8]$.

20.

$$f(X) = X^4 + \underbrace{8X^3 - 3X^2} - 130X - 200$$

One positive zero

$$f(-X) =$$
$$(-X)^4 + 8(-X)^3 - 3(-X)^2 - 130(-X) - 200$$
$$= \underbrace{X^4 - 8X^3} - 3X^2 + \underbrace{130X - 200}$$

Three or one negative zeros

21.

$$f(X) = \underbrace{X^7 - 3X^5} - \underbrace{4X^4 + 2X} + 7$$

Two or zero positive zeros

$$f(-X) =$$
$$(-X)^7 - 3(-X)^5 - 4(-X)^4 + 2(-X) + 7$$
$$= \underbrace{-X^7 + 3X^5} - \underbrace{4X^4 - 2X} + 7$$

Three or one negative zeros

22.

$$f(X) = \underbrace{-7X^5 + 7X^3} - \underbrace{7X^2 + 5X} + 1$$

Three or one positive zeros

$$f(-X) =$$
$$-7(-X)^5 + 7(-X)^3 - 7(-X)^2 + 5(-X) + 1$$
$$= \underbrace{7X^5 - 7X^3} - 7X^2 - \underbrace{5X + 1}$$

Two or zero negative zeros

23.

$$f(X) = \underbrace{3X^4 - 3X^3} + \underbrace{2X^2 + 10X} + 4$$

Two or zero positive zeros

$$f(-X) =$$
$$3(-X)^4 - 3(-X)^3 + 2(-X)^2 + 10(-X) + 4$$
$$= 3X^4 + \underbrace{3X^3 + 2X^2} - \underbrace{10X + 4}$$

Two or zero negative zeros

24.

$$X^3 + 13X^2 + 7X - 165 = 0$$

$$\pm 1 \qquad \begin{array}{l} \pm 1, \pm 3, \pm 5, \pm 11, \pm 15, \\ \pm 33, \pm 55, \pm 165 \end{array}$$

Possible zeros:
$$\pm 1, \pm 3, \pm 5, \pm 11, \pm 15, \pm 33 \pm 55, \pm 165$$

$$
\begin{array}{r|rrrr}
3 & 1 & 13 & 7 & -165 \\
  &   & 3 & 48 & 165 \\
\hline
  & 1 & 16 & 55 & 0
\end{array}
$$

$$(X - 3)(X^2 + 16X + 55) =$$
$$(X - 3)(X + 5)(X + 11)$$
$$X = -11, \ X = -5, \text{ and } X = 3$$

25.

$$X^3 + 4X^2 - 4X - 16 = 0$$
$$\pm 1 \qquad \pm 1, \pm 2, \pm 4, \pm 8, \pm 16$$

Possible zeros: $\pm 1, \pm 2, \pm 4, \pm 8, \pm 16$

$$
\begin{array}{r|rrrr}
2 & 1 & 4 & -4 & -16 \\
  &   & 2 & 12 & 16 \\
\hline
  & 1 & 6 & 8 & 0
\end{array}
$$

$$(X - 2)(X^2 + 6X + 8)$$
$$(X - 2)(X + 4)(X + 2)$$
$$X = -4, \ X = -2, \text{ and } X = 2$$

26.

$$X^3 - 6X^2 - X + 30 = 0$$

± 1 $\pm 1, \pm 2, \pm 3, \pm 5, \pm 6,$

             $\pm 10, \pm 15, \pm 30$

Possible zeros: $\pm 1, \pm 2, \pm 3, \pm 5, \pm 6,$
$\pm 10, \pm 15, \pm 30$

$$\begin{array}{r|rrrr} 5 & 1 & -6 & -1 & 30 \\ & & 5 & -5 & -30 \\ \hline & 1 & -1 & -6 & 0 \end{array}$$

$(X - 5)(X^2 - X - 6)$
$(X - 5)(X - 3)(X + 2)$
     $X = -2$, $X = 3$, and $X = 5$

27.

$$2X^3 + X^2 - 26X - 40 = 0$$

$\pm 1, \pm 2$ $\pm 1, \pm 2 \pm 4, \pm 5, \pm 8,$

              $\pm 10, \pm 20, \pm 40$

Possible zeros: $\pm \dfrac{1}{2} \pm 1, \pm 2, \pm \dfrac{5}{2}, \pm 4,$
$\pm 5, \pm 8, \pm 10, \pm 20, \pm 40$

$$\begin{array}{r|rrrr} 4 & 2 & 1 & -26 & -40 \\ & & 8 & 36 & 40 \\ \hline & 2 & 9 & 10 & 0 \end{array}$$

$(X - 4)(2X^2 + 9X + 10)$
$(X - 4)(X + 2)(2X + 5)$
     $X = -\dfrac{5}{2}$, $X = -2$, and $X = 4$

28.

$$X^3 + 16X^2 - 73X - 952 = 0$$

± 1 $\pm 1, \pm 2, \pm 4, \pm 7, \pm 8,$
$\pm 14, \pm 17, \pm 28, \pm 34,$
$\pm 56, \pm 119, \pm 238,$
$\pm 476, \pm 952$

Possible zeros: $\pm 1, \pm 2, \pm 4, \pm 7, \pm 8, \pm 14, \pm 17,$
$\pm 28, \pm 34, \pm 56, \pm 119, \pm 238, \pm 476, \pm 952$

$$\begin{array}{r|rrrr} -7 & 1 & 16 & -73 & -952 \\ & & -7 & -63 & 952 \\ \hline & 1 & 9 & -136 & 0 \end{array}$$

$(X + 7)(X^2 + 9X - 136)$
$(X + 7)(X - 8)(X + 17)$
     $X = -17$, $X = -7$, and $X = 8$

29.

$$X^4 + 5X^3 - 30X^2 - 40X + 64 = 0$$

$$\pm 1 \qquad \begin{matrix} \pm 1, \pm 2, \pm 4, \pm 8, \\ \pm 16, \pm 32, \pm 64 \end{matrix}$$

Possible zeros:

$$\pm 1, \pm 2, \pm 4, \pm 8, \pm 16, \pm 32, \pm 64$$

$$\begin{array}{r|rrrrr} -2 & 1 & 5 & -30 & -40 & 64 \\ & & -2 & -6 & 72 & -64 \\ \hline & 1 & 3 & -36 & 32 & 0 \end{array}$$

$$(X + 2)(X^3 + 3X^2 - 36X + 32)$$

Factor $X^3 + 3X^2 - 36X + 32$

$$\pm 1 \qquad \begin{matrix} \pm 1, \pm 2, \pm 4, \pm 8, \\ \pm 16, \pm 32 \end{matrix}$$

Possible zeros: $\pm 1, \pm 2, \pm 4, \pm 8, \pm 16, \pm 32$

$$\begin{array}{r|rrrr} 1 & 1 & 3 & -36 & 32 \\ & & 1 & 4 & -32 \\ \hline & 1 & 4 & -32 & 0 \end{array}$$

$$(X - 1)(X^2 + 4X - 32)$$
$$(X - 1)(X - 4)(X + 8)$$

$$(X + 2)(X - 1)(X - 4)(X + 8) = 0$$
$$X = -8, \ X = -2, \ X = 1, \text{ and } X = 4$$

30.

$$X^4 - 4X^3 - 22X^2 + 4X + 21 = 0$$
$$\pm 1 \qquad \pm 1, \pm 3, \pm 7. \pm 21$$

Possible zeros: $\pm 1, \pm 3, \pm 7. \pm 21$

$$\begin{array}{r|rrrrr} 1 & 1 & -4 & -22 & 4 & 21 \\ & & 1 & -3 & -25 & -21 \\ \hline & 1 & -3 & -25 & -21 & 0 \end{array}$$

$$(X - 1)(X^3 - 3X^2 - 25X - 21)$$

Factor $X^3 - 3X^2 - 25X - 21$
$$\pm 1 \qquad \pm 1, \pm 3, \pm 7, \pm 21$$

Possible zeros: $\pm 1, \pm 3, \pm 7, \pm 21$

$$\begin{array}{r|rrrr} -3 & 1 & -3 & -25 & -21 \\ & & -3 & 18 & 21 \\ \hline & 1 & -6 & -7 & 0 \end{array}$$

$$(X + 3)(X^2 - 6X - 7)$$
$$(X + 3)(X + 1)(X - 7)$$

$$(X - 1)(X + 3)(X + 1)(X - 7) = 0$$
$$X = -3, \ X = -1, \ X = 1, \text{ and } X = 7$$

Chapter 5   Graphing

▶ We can write a linear equation in slope-intercept form: $Y = mX + b$. The "m" is the slope, which indicates the steepness of the graph. The slope is rise over run; we can find the slope if we have two points on the line by applying the formula $m = \dfrac{y_2 - y_1}{x_2 - x_1}$. The "$b$" is the y-intercept, which is the point where the graph crosses the y-axis.

Example 1:
Find the equation of line
(slope-intercept form).

$$\text{Slope} = 2, \text{ goes through } (-1,5)$$

$Y = mX + b$ — We apply the slope-intercept formula. We have the slope ($m = 2$). We find the y-intercept by substituting the corresponding values into the slope-intercept equation.

$5 = (2)(-1) + b$ — We substitute the y-coordinate into Y, x-coordinate into X, and the slope into m.

$5 = -2 + b$ — We solve for the value of b.

$b = 7$

$Y = 2X + 7$ — We have the slope-intercept form.

Example 2:
Find the equation of line
(slope-intercept form).

$(-3,5), (2,25)$ — We have two points on the line.

$m = \dfrac{y_2 - y_1}{x_2 - x_1}$ — We first find the slope.

$(x_1, y_1), (x_2, y_2)$
$(-3,5), (2,25)$

$m = \dfrac{25 - 5}{2 - (-3)} = \dfrac{20}{5} = 4$ — We substitute the corresponding values.

$m = 4$, $(2,25)$

Once we find the slope, we pick one of the points on the line. Then, we substitute into the slope-intercept form to find the y-intercept.

$25 = 4(2) + b$

Notice that this process is same as that of example 1.

$b = 17$

$Y = 4X + 17$

» When we graph the inequality, we graph the equation first and determine which region to shade. The inequalities $<$ and $>$ make the graphs dotted lines. The inequalities $\leq$ and $\geq$ make the graphs solid lines.

$$Y \geq X^2$$

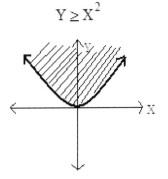

$$Y > X^2$$

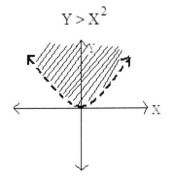

For $\geq$, the graph is solid. We shade the region above the graph because we have "Y is greater than or equal to".

For $>$, the graph is dotted. We shade the region above the graph because we have "Y is greater than".

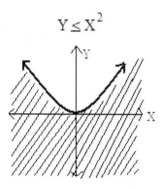

$Y \leq X^2$

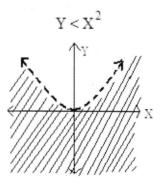

$Y < X^2$

For ≤, the graph is solid. We shade the region below the graph because we have "Y is less than or equal to".

For <, the graph is dotted. We shade the region below the graph because we have "Y is less than".

Example 3:
Graph the inequality.

$Y < -X^2 + 2$

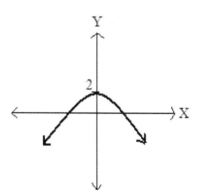

We graph the equation $Y = -X^2 + 2$ first.

Because of the inequality <, we make the graph line dotted.

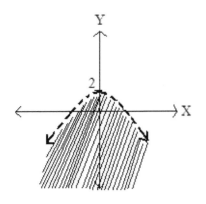

We shade the region below the graph because we have "Y is less than" an equation.

Now, we have graphed our equation

$$Y < -X^2 + 2$$

» The cubic function is $Y = X^3$.

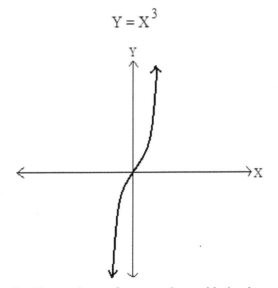

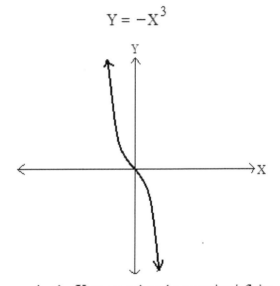

In this graph, we focus on the end behaviors. As the X approaches negative infinity, the Y approaches negative infinity. As the X approaches positive infinity, the Y approaches positive infinity.

As the X approaches the negative infinity, the Y approaches positive infinity.
As the X approaches the positive infinity, the Y approaches negative infinity.

The end behaviors of cubic function will remain the same regardless of the equation. For cubic functions like $Y = X^3 - 2X^2 + X - 1$, the end behaviors will be the same as that of $Y = X^3$.

Having studied the end behaviors, we know how the graph will look like if we know the zeros of the function.

Example 4:
Sketch the graph.

$$Y = X^3 - 5X^2 - 4X + 20$$
X-intercepts: $(-2,0)$, $(2,0)$, $(5,0)$

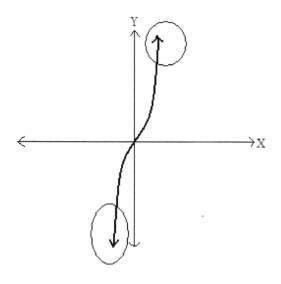

The graph will have the same end behaviors as that of the graph $Y = X^3$. The circled parts of the graph $Y = X^3$ are the end behaviors. As X approaches negative infinity, the Y approaches negative infinity. As X approaches positive infinity, the Y approaches positive infinity.

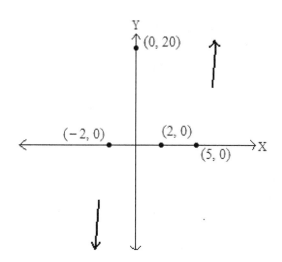

We identify the intercepts. The X-intercepts were given. We draw dots for the X-intercepts. We find the Y-intercept by substituting 0 for the X in the original equation and evaluating for the value of Y. The Y-intercept in this graph is $(0, 20)$. We connect the intercepts to draw the graph.

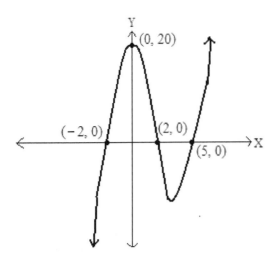

X-intercept has multiplicity. The multiplicity means how many times the zero is a solution. In this example, each X-intercept has odd multiplicity (multiplicity of 1), so the graph crosses the intercepts. If the intercept has even multiplicity, the graph will bounce (reflect) at the intercept rather than cross the intercept.

» Finding and drawing asymptotes will help in graphing. To find vertical asymptote, we set the denominator equal to 0 and solve for X, which will be the vertical asymptote. A fraction with 0 in the denominator is undefined $\left(\dfrac{number}{0} = undefined \right)$. We may have vertical asymptotes X=number when we have denominator. We will set the denominator equal to 0 when we see a denominator.

We also have horizontal asymptotes. Instead of vertical asymptote, the horizontal asymptote is a horizontal line (Y=number). To find a horizontal asymptote, we compare the degree of the numerator and the denominator. If the degree of the numerator is less than the degree of the denominator, we have a horizontal asymptote at $Y = 0$. If the degree of the numerator is equal to the degree of the denominator, we have horizontal asymptote equal to the leading coefficient of the numerator divided by the leading coefficient of the denominator.

If the numerator degree is greater than the denominator, we do not have a horizontal asymptote. However, we have a third type of asymptote called the oblique or slant asymptote. To find the oblique or slant asymptote, we verify that the numerator degree is greater than the denominator degree by 1. Then, we divide the fraction and the quotient is the oblique or slant asymptote (disregard the remainder).

Example 5:
Find the asymptotes (vertical, horizontal or/ and oblique asymptotes).

$$f(X) = \frac{2X^2 - 1}{X^2 + 7X + 10}$$

$X^2 + 7X + 10 = 0$ We set the denominator equal to 0 to find
$(X + 2)(X + 5) = 0$ vertical asymptote(s). We solve for X.
$X = -2$ and $X = -5$ We have two vertical asymptotes.

Numerator degree = denominator degree
Horizontal asymptote:

$$Y = \frac{2}{1} = 2$$

We see that the numerator degree equals the denominator degree. We divide the numerator's leading coefficient by the denominator's leading coefficient to obtain the horizontal asymptote.

Example 6:
Find the asymptotes (vertical, horizontal, or/ and oblique asymptotes).

$$f(X) = \frac{3X^3 + 2X - 1}{X^2 - 4}$$

$$X^2 - 4 = 0$$
$$(X - 2)(X + 2) = 0$$
$$X - 2 = 0 \quad \text{or} \quad X + 2 = 0$$
$$X = 2 \quad \text{or} \quad X = -2$$

We set the denominator equal to 0. We solve for the values of X (the vertical asymptotes).

Numerator degree > denominator degree
We do not have horizontal asymptote, but we have an oblique asymptote.

$$X^2 - 4 \overline{)3X^3 + 2X - 1} \quad \overset{3X}{}$$

$$Y = 3X$$

We find the quotient, which is 3X, and we do not worry about the remainder. The quotient is the oblique asymptote.

» We can graph the asymptotes first. Then, we can apply the point-plotting method to see how the graph looks like in certain parts of the domain. To apply the point-plotting method, we substitute the values of x-coordinates into the equation to find the corresponding y-coordinates. Then, we plot the coordinates on the graph and connect the dots.

Example 7:

Sketch the graph.

$$f(X) = \frac{2X^2 - 5}{X^2 + X - 2}$$

$X^2 + X - 2 = 0$ We set the denominator equal to 0 to

$(X + 2)(X - 1) = 0$ find the vertical asymptotes.

$X = -2$ and $X = 1$

Numerator degree = denominator degree

$Y = \dfrac{2}{1} = 2$ We divide the leading coefficient of

the numerator by the leading

$Y = 2$ coefficient of the denominator. The

horizontal asymptote is $Y = 2$.

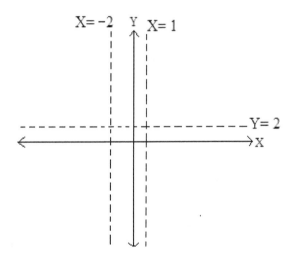

We first draw the asymptotes (vertical, horizontal, or/ and oblique asymptotes).

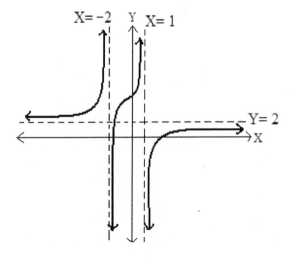

We apply the point-plotting method to sketch how the graph looks in certain parts of the domain. The graph approaches the asymptotes but does not touch the asymptotes, although the graph can cross the asymptote as it does in one of the points in the graph on the left. The more x-coordinates we substitute into the equation to find the corresponding y-coordinates, the more accurate the graph will be.

Chapter 5   Graphing

1. Find the equation of line (slope-intercept form). Slope=5, goes through $(1,-4)$	2. Find the equation of line (slope-intercept form). Slope= -2, goes through $(0,5)$
3. Find the equation of line (slope-intercept form). $(1,2)$ and $(-2,-3)$	4. Find the equation of line (slope-intercept form). $(-4,2)$ and $(3,-5)$
5. Graph the inequality. $Y \leq -X + 7$	6. Graph the inequality. $Y \geq X^2 + 4X + 3$
7. Graph the inequality. $Y < -X^2 + 5$	8. Graph the inequality. $Y > -X^4 - 1$
9. Sketch the graph. $Y = X^3 - X^2 - 10X - 8$ X-intercepts: $(-2,0), (-1,0), (4,0)$	10. Sketch the graph. $Y = X^3 + 4X^2 - 7X - 10$ X-intercepts: $(-5,0), (-1,0), (2,0)$

Chapter 5   Graphing

11. Sketch the graph.	12. Sketch the graph.
$Y = -X^3 - 3X^2 + 6X + 8$	$Y = X^3 + X^2 - 10X + 8$
X-intercepts: $(-4,0)(-1,0)(2,0)$	X-intercepts: $(-4,0)(1,0)(2,0)$
13. Sketch the graph.	14. Sketch the graph.
$Y = 2X^3 - 5X^2 - 14X + 8$	$Y = -3X^3 - 25X^2 - 29X - 7$
X-intercepts: $(-2,0)\left(\frac{1}{2},0\right),(4,0)$	X-intercepts: $(-7,0)(-1,0)\left(\frac{-1}{3},0\right)$
15. Sketch the graph.	16. Sketch the graph.
$Y = 2X^3 + 11X^2 + 10X - 8$	$Y = -6X^3 + 25X^2 - 3X - 4$
X-intercepts: $(-4,0)(-2,0)\left(\frac{1}{2},0\right)$	X-intercepts: $\left(\frac{-1}{3},0\right),\left(\frac{1}{2},0\right),(4,0)$
17. Find the asymptotes (vertical, horizontal, or/and oblique asymptotes).	18. Find the asymptotes (vertical, horizontal, or/and oblique asymptotes).
$f(X) = \dfrac{5X^2 + 1}{X^2 + 3X - 10}$	$f(X) = \dfrac{X + 5}{2X^2 + X - 3}$
19. Find the asymptotes (vertical, horizontal, or/and oblique asymptotes).	20. Find the asymptotes (vertical, horizontal, or/and oblique asymptotes).
$f(X) = \dfrac{X^2 + X + 2}{X + 2}$	$f(X) = \dfrac{X + 2}{X^2 - 3X - 4}$

Chapter 5   Graphing

21. Find the asymptotes (vertical, horizontal, or/and oblique asymptotes). $$f(X)=\frac{4X^2-4X+1}{3X^2+1}$$	22. Find the asymptotes (vertical, horizontal, or/and oblique asymptotes). $$f(X)=\frac{X+7}{X^2-2X-3}$$
23. Find the asymptotes (vertical, horizontal, or/and oblique asymptotes). $$f(X)=\frac{5X^2+3}{X^2+2X-8}$$	24. Find the asymptotes (vertical, horizontal, or/and oblique asymptotes). $$f(X)=\frac{4X^2+1}{X+2}$$
25. Sketch the graph. $$f(X)=\frac{X+5}{X^2+3X-4}$$	26. Sketch the graph. $$f(X)=\frac{X-4}{X^2+2X-15}$$
27. Sketch the graph. $$f(X)=\frac{X^2+X-2}{X^2-X-12}$$	28. Sketch the graph. $$f(X)=\frac{X^2+1}{X-1}$$
29. Sketch the graph. $$f(X)=\frac{2X^2+5}{X^2+2X-3}$$	30. Sketch the graph. $$f(X)=\frac{2X^2+3}{2X-1}$$

Chapter 5 Graphing
Answer key

1. $y = mx + b$
   $-4 = 5(1) + b$
   $b = -9$
   $Y = 5X - 9$

2. $y = mx + b$
   $5 = -2(0) + b$
   $b = 5$
   $Y = -2X + 5$

3. $m = \dfrac{-3-2}{-2-1} = \dfrac{-5}{-3} = \dfrac{5}{3}$
   $y = mx + b$
   $2 = \dfrac{5}{3}(1) + b$
   $b = \dfrac{1}{3}$
   $Y = \dfrac{5}{3}X + \dfrac{1}{3}$

4. $m = \dfrac{-5-2}{3-(-4)} = \dfrac{-7}{7} = -1$
   $y = mx + b$
   $-5 = (-1)(3) + b$
   $b = -2$
   $Y = -X - 2$

5.

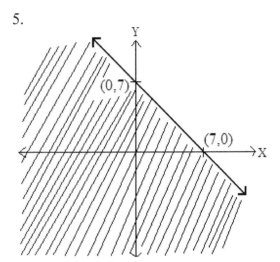

6.

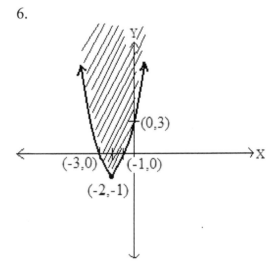

7.

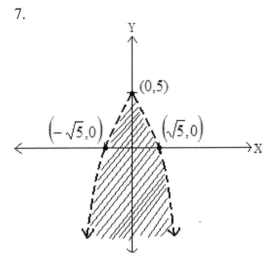

8.

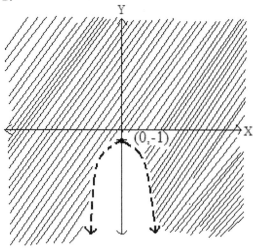

11.

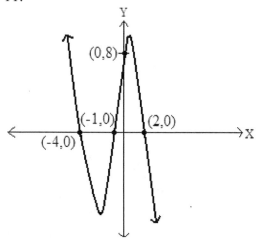

9.

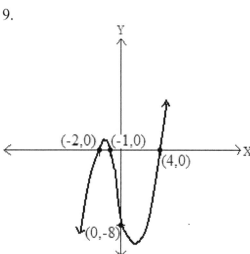

12.

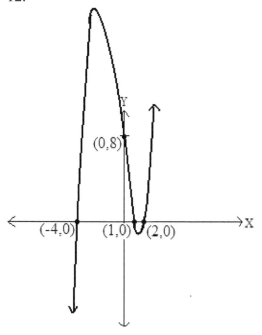

10.

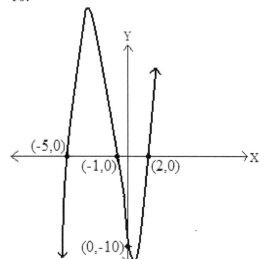

13.

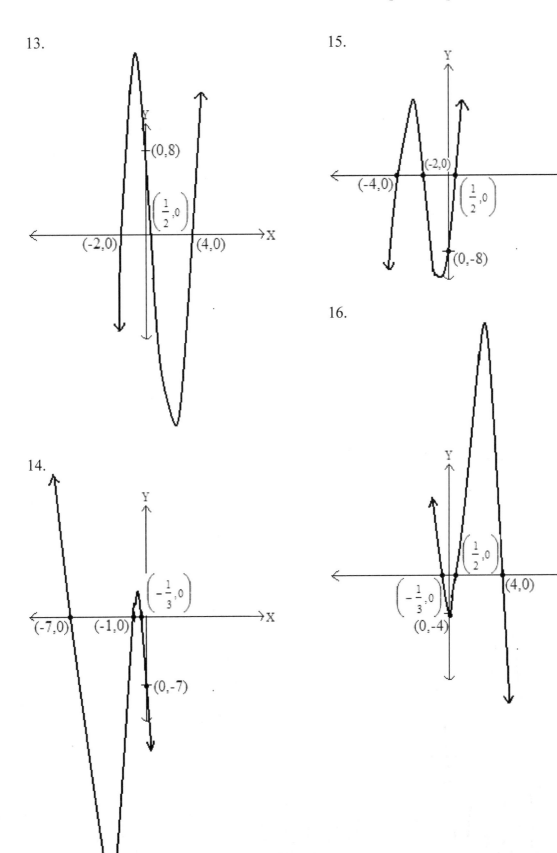

15.

16.

14.

17. $\dfrac{5X^2+1}{X^2+3X-10} = \dfrac{5X^2+1}{(X-2)(X+5)}$

$X-2=0$ and $X+5=0$

Vertical asymptotes: $X=-5$ and $X=2$

Numerator degree=denominator degree;

H.A.=$\dfrac{\text{leading coefficient of numerator}}{\text{leading coefficient of denominator}}$

Horizontal asymptote: $Y=\dfrac{5}{1}=5$

18. $\dfrac{X+5}{2X^2+X-3} = \dfrac{X+5}{(X-1)(2X+3)}$

$X-1=0$ and $2X+3=0$

Vertical asymptotes: $X=\dfrac{-3}{2}$ and $X=1$

Numerator degree < denominator degree

H.A.=0

Horizontal asymptote: $Y=0$

19. $X+2=0$

Vertical asymptote: $X=-2$

Numerator degree > denominator degree

Oblique asymptote

$(X^2+X+2)\div(X+2)=X-1+\dfrac{4}{X+2}$

Oblique asymptote: $Y=X-1$

20. $\dfrac{X+2}{X^2-3X-4} = \dfrac{X+2}{(X-4)(X+1)}$

$X-4=0$ and $X+1=0$

Vertical asymptotes: $X=4$ and $X=-1$

Numerator degree < denominator degree

H.A.=0

Horizontal asymptote: $Y=0$

21. Vertical asymptotes: N/A

Numerator degree = denominator degree

H.A.=$\dfrac{\text{leading coefficient of numerator}}{\text{leading coefficient of denominator}}$

Horizontal asymptote: $Y=\dfrac{4}{3}$

22. $\dfrac{X+7}{X^2-2X-3} = \dfrac{X+7}{(X-3)(X+1)}$

$X-3=0$ and $X+1=0$

Vertical asymptotes: $X=3$ and $X=-1$

Numerator degree < denominator degree

H.A.=0

Horizontal asymptote: $Y=0$

23. $\dfrac{5X^2+3}{X^2+2X-8} = \dfrac{5X^2+3}{(X+4)(X-2)}$

$X+4=0$ and $X-2=0$

Vertical asymptotes: $X=-4$ and $X=2$

Numerator degree = denominator degree

H.A.=$\dfrac{\text{leading coefficient of numerator}}{\text{leading coefficient of denominator}}$

Horizontal asymptote: $Y=\dfrac{5}{1}=5$

24. Vertical asymptotes: $X=-2$

Numerator degree > denominator degree

Oblique asymptote

$(4X^2+1)\div(X+2)=4X-8+\dfrac{17}{X+2}$

Oblique asymptote: $Y=4X-8$

25. We find and draw the asymptotes and apply the point-plotting method.

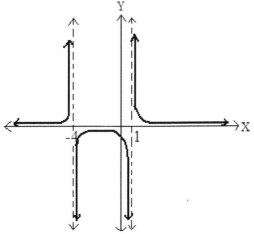

28. We find and draw the asymptotes and apply the point-plotting method.

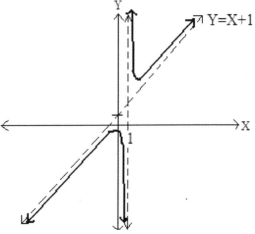

26. We find and draw the asymptotes and apply the point-plotting method.

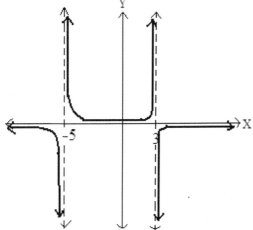

29. We find and draw the asymptotes and apply the point-plotting method.

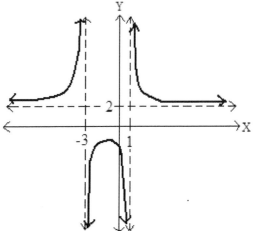

27. We find and draw the asymptotes and apply the point-plotting method.

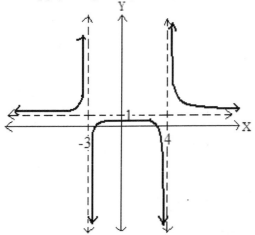

30. We find and draw the asymptotes and apply the point-plotting method.

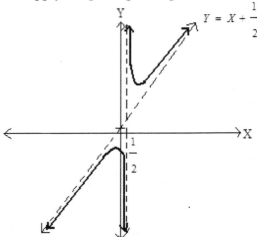

Chapter 6   The systems of non-linear equations

▶ We will study the systems of non-linear equations in which we have an equation or equations that are not lines. We will study two methods to solve: substitution method and elimination method.

In substitution method, we set one equation equal to one variable, and then we substitute that variable into the other equation.

Example 1:

Find the solutions.

$$-5X^2 + 10Y^2 = 5$$
$$-X + 2Y = 3$$

We are going to set one equation equal to a variable. The second equation $-X + 2Y = 3$ looks easier to set it equal to one variable.

$$-X + 2Y = 3$$
$$-X = 3 - 2Y$$
$$X = 2Y - 3$$

$$-5X^2 + 10Y^2 = 5$$

We substitute second equation into the first equation.

$$\uparrow$$

$$X = 2Y - 3$$

$$-5(2Y - 3)^2 + 10Y^2 = 5$$
$$-5(4Y^2 - 12Y + 9) + 10Y^2 = 5$$
$$-20Y^2 + 60Y - 45 + 10Y^2 = 5$$
$$-10Y^2 + 60Y - 50 = 0$$
$$-10(Y - 5)(Y - 1) = 0$$
$$(Y - 5)(Y - 1) = 0$$
$$Y = 5 \text{ and } Y = 1$$

Now, we have everything in terms of Y. We solve for Y.

$$-X + 2Y = 3$$
$$-X + 2(5) = 3$$
$$-X + 10 = 3$$
$$X = 7$$
$$(X, Y) \rightarrow (7, 5)$$

We have $Y = 5$ and $Y = 1$. We substitute the y-value in the original equation to find the value of corresponding X.

$$-X + 2(1) = 3$$
$$-X = 1$$

$$X = -1$$
$$(X,Y) \rightarrow (-1,1)$$

Our solutions are $(7,5)$ and $(-1,1)$.

» Now, we will study the elimination method. In most cases, we cannot apply the elimination method when one equation is a linear equation and the other is a non-linear equation. We can apply the elimination method if both equations are non-linear.

Example 2:

$$-2X^2 + 5Y^2 = 30$$
$$X^2 - 2Y^2 = -7$$

We try to eliminate either X^2 or Y^2.

$$-2X^2 + 5Y^2 = 30$$
$$2(X^2 - 2Y^2 = -7)$$

We will eliminate X^2. We multiply the second equation by 2 to make X^2 opposites.

$$-2X^2 + 5Y^2 = 30$$
$$2X^2 - 4Y^2 = -14$$

$$Y^2 = 16$$
$$Y = \pm 4$$
$$Y = 4, \ Y = -4$$

$$X^2 - 2Y^2 = -7$$
$$X^2 - 2(4)^2 = -7$$
$$X^2 - 32 = -7$$
$$X^2 = 25$$
$$X = \pm 5$$
$$(X,Y) \rightarrow (5,4), \ (-5,4)$$

Once we solve one variable, we substitute the value into the original equation to solve the other variable. Notice that this step is the same as that of the substitution method.

$$X^2 - 2(-4)^2 = -7$$
$$X^2 - 32 = -7$$
$$X^2 = 25$$
$$X = \pm 5$$
$$(X,Y) \rightarrow (-5,-4), \ (5,-4)$$

Our solutions are $(5,4)$, $(-5,4)$, $(-5,-4)$, and $(5,-4)$

» We can also find the solutions by graphing the equations. The intersections of the two graphs are the solutions.

Chapter 6   The systems of non-linear equations

1. Find the solutions. $X^2 - 2Y^2 = 2$ $-X + 2Y = 4$	2. Find the solutions. $2X^2 + 3Y^2 = 77$ $3Y = 4X + 11$
3. Find the solutions. $4X^2 - Y = 21$ $4X + Y = 3$	4. Find the solutions. $10X^2 - 7Y^2 = 3$ $X + Y = 0$
5. Find the solutions. $3X^2 + Y = 20$ $3X + Y = 14$	6. Find the solutions. $6X^2 - Y^2 = 38$ $2X - 2 = Y$
7. Find the solutions. $Y - 2X^2 = -25$ $Y + 9 = -4X$	8. Find the solutions. $X^2 + 3Y^2 = 148$ $Y + X = 6$
9. Find the solutions. $X + 4Y^2 = 102$ $18 = 4Y - X$	10. Find the solutions. $4X^2 - 5Y^2 = -316$ $-8X = Y$

Chapter 6   The systems of non-linear equations

11. Find the solutions.	12. Find the solutions.
$5X^2 - 4Y^2 = -59$ $-5X + 4Y = 11$	$X^2 + 6Y^2 = 175$ $2Y + 5 = -X$
13. Find the solutions.	14. Find the solutions.
$X + 5Y^2 = 127$ $-5Y - 23 = X$	$-7X^2 + Y^2 = -3$ $\dfrac{7}{3}X + Y = \dfrac{-1}{3}$
15. Find the solutions.	16. Find the solutions.
$-X^2 - Y^2 = -101$ $-X + Y = 9$	$X^2 + 5Y^2 = 21$ $Y + 3 = -X$
17. Find the solutions.	18. Find the solutions.
$X - 4Y^2 = 4$ $X + 4Y = 12$	$7X^2 - 2Y^2 = -22$ $2Y = 5X$
19. Find the solutions.	20. Find the solutions.
$3X^2 + 10Y^2 = 43$ $Y = -2X$	$8X^2 - 15Y^2 = 8$ $-11Y - 8X = 0$

Chapter 6   The systems of non-linear equations

21. Find the solutions. $X^2 + 3Y^2 = 31$ $2X^2 - Y^2 = -1$	22. Find the solutions. $X^2 - 5Y^2 = -121$ $2X^2 + Y^2 = 33$
23. Find the solutions. $5X^2 - Y^2 = -11$ $3X^2 + 4Y^2 = 67$	24. Find the solutions. $7X^2 + Y^2 = 32$ $X^2 - 2Y^2 = -4$
25. Find the solutions. $X^2 + 10Y^2 = 65$ $-4X^2 - Y^2 = -104$	26. Find the solutions. $8X^2 - Y^2 = 71$ $2X^2 + 5Y^2 = 23$
27. Find the solutions. $-9X^2 - Y^2 = -109$ $X^2 + 5Y^2 = 501$	28. Find the solutions. $3X^2 - Y^2 = -46$ $2X^2 + 3Y^2 = 413$
29. Find the solutions. $11X^2 - 5Y^2 = -1$ $14X^2 + 4Y^2 = 92$	30. Find the solutions. $4X^2 + Y^2 = 325$ $-12X^2 + 5Y^2 = 825$

Chapter 6 The systems of non-linear equations
Answer key

1. $(-2,1),(10,7)$

2. $(1,5),(-5,-3)$

3. $(2,-5),(-3,15)$

4. $(1,-1),(-1,1)$

5. $(2,8),(-1,17)$

6. $(3,4),(-7,-16)$

7. $(-4,7),(2,-17)$

8. $(-1,7),(10,-4)$

9. $(2,5),(-42,-6)$

10. $(1,-8),(-1,8)$

11. $(1,4),(-23,-26)$

12. $(5,-5),(-11,3)$

13. $(2,-5),(-53,6)$

14. $(-1,2),(2,-5)$

15. $(1,10),(-10,-1)$

16. $(-4,1),(-1,-2)$

17. $(8,1),(20,-2)$

18. $(2,5),(-2,-5)$

19. $(-1,2),(1,-2)$

20. $(-11,8),(11,-8)$

21. $(-2,-3),(-2,3),(2,-3),(2,3)$

22. $(-2,-5),(-2,5),(2,-5),(2,5)$

23. $(-1,-4),(-1,4),(1,-4),(1,4)$

24. $(-2,-2),(-2,2),(2,-2),(2,2)$

25. $(-5,-2),(-5,2),(5,-2),(5,2)$

26. $(-3,-1),(-3,1),(3,-1),(3,1)$

27. $(-1,-10),(-1,10),(1,-10),(1,10)$

28. $(-5,-11),(-5,11),(5,-11),(5,11)$

29. $(-2,-3),(-2,3),(2,-3),(2,3)$

30. $(-5,-15),(-5,15),(5,-15),(5,15)$

Chapter 7   Sequence and series

▶ Sequence is a certain number in a sequence of numbers. We may try to find the 12th number in a sequence. Series, on the other hand, is a sum of numbers.

We will study arithmetic sequence. A sequence is arithmetic if the number increases by same constant number.

$$5 \xrightarrow{+7} 12 \xrightarrow{+7} 19 \xrightarrow{+7} 26 \xrightarrow{+7} 33 \xrightarrow{+7} 40 \ldots$$

We can keep adding a constant number to find the sequence. However, it would be time consuming to find a number in a sequence that is very far away. We have an arithmetic sequence formula to help us find the number.

$$a_n = a_1 + (n-1)d$$

$a_n = n^{th}$ term in the sequence

$a_1 = $ first term

$n = $ number of numbers

$d = $ difference (this is same as the number that increases constantly)

Difference is subtraction, but it is easier to think about adding a constant number.

Example 1:
Find the 52nd term.

8, 12, 16, 20, …

$8 \xrightarrow{+4} 12 \xrightarrow{+4} 16 \xrightarrow{+4} 20$ We have an arithmetic sequence.

$a_n = a_1 + (n-1)d$ We apply the arithmetic sequence formula to find a_{52}.

$a_1 = $ first term $= 8$

$n = $ number of numbers $= 52$

$d = $ difference $= 4$

$a_{52} = 8 + (52-1)(4)$

$a_{52} = 212$ 52nd term is 212.

» We also have a special sequence called the geometric sequence. The number in the sequence is multiplied by a constant number.

5, 15, 45, 135, 405, …

$$5 \xrightarrow{\times 3} 15 \xrightarrow{\times 3} 45 \xrightarrow{\times 3} 135 \xrightarrow{\times 3} 405$$

We have a geometric sequence formula that helps us to find the nth term in the sequence.

$$a_n = a_1 r^{n-1}$$

$a_n = n^{th}$ term
$a_1 =$ first term
$r =$ ratio (or the constant number multiplied)
$n =$ number of numbers

Ratio is division, but it is easier to think about ratio as a constant number multiplied.

$$5 \xrightarrow{\frac{15}{5}=3} 15 \xrightarrow{\frac{45}{15}=3} 45 \xrightarrow{\frac{135}{45}=3} 135 \xrightarrow{\frac{405}{135}=3} 405$$

$$5 \xrightarrow{\times 3} 15 \xrightarrow{\times 3} 45 \xrightarrow{\times 3} 135 \xrightarrow{\times 3} 405$$

Example 2:
Find the 8th term.

1, 4, 16, 64, …

$1 \xrightarrow{\times 4} 4 \xrightarrow{\times 4} 16 \xrightarrow{\times 4} 64$ We have a geometric sequence.

$a_n = a_1 r^{n-1}$ We apply the geometric sequence formula to find the 8th term, or a_8.

$a_1 =$ first term $= 1$
$r =$ ratio $= 4$
$n =$ number of numbers $= 8$

$$a_8 = 1(4)^{8-1}$$
$$a_8 = 16,384$$

» Series is a sum of numbers in a sequence. For the sum of arithmetic sequence, we apply the arithmetic sum formula.

$$S_n = \frac{n}{2}\left(first.term + last.term\right) = \frac{n}{2}\left(a_1 + a_n\right)$$

S_n = sum of sequence of n numbers

a_1 = first term

a_n = last term

Example 3:

Find the sum of first 65 terms.

7, 12, 17, 22, …   We have an arithmetic sequence.

$$S_n = \frac{n}{2}\left(first.term + last.term\right) = \frac{n}{2}\left(a_1 + a_n\right)$$ We apply the arithmetic sum formula.

a_1 = first term = 7

n = number of terms = 65

last term = $a_n = a_1 + (n-1)d$ We apply the arithmetic sequence

$\qquad\qquad a_{65} = 7 + (65-1)(5)$ formula to find the last term.

$\qquad\qquad a_{65} = 327$

$$S_{65} = \frac{65}{2}(7+327)$$

$$S_{65} = 10{,}855$$

» For the sum of geometric sequence, we apply the geometric sum formula.

$$S_n = a_1\left(\frac{1-r^n}{1-r}\right)$$

S_n = sum of n numbers in geometric sequence

a_1 = first term

r = ratio

n = number of numbers

Example 4:

Find the sum of first 15 terms.

3, 6, 12, 24, …   We have a geometric sequence.

$$S_n = a_1 \left(\frac{1 - r^n}{1 - r} \right)$$ We apply the geometric sum formula.

$a_1 =$ first term $= 3$
$r =$ ratio $= 2$
$n =$ number of numbers $= 15$

$$S_{15} = 3 \left(\frac{1 - 2^{15}}{1 - 2} \right)$$

$$S_{15} = 98,301$$

» Sigma $\left(\sum\limits_{t=}^{n} \right)$ is a notation for sum.

$\sum\limits_{t=1}^{4} t$ Here, t=1 means we start from the first term. 4 indicates 4 terms.

first   second third fourth
term   term   term term
$\ \ 1 \ + \ 2 \ + \ 3 \ + \ 4 \ \ \ \ = 10$

We also have a formula for a variable to the first power.

$$\sum\limits_{t=1}^{4} t = \frac{n(n+1)}{2} = \frac{4(5)}{2} = 10$$

We can utilize the properties of sigma.

$$\sum\limits_{t=1}^{n} at = a \sum\limits_{t=1}^{n} t \qquad\qquad (a = number)$$

$$\sum\limits_{t=1}^{n} (t + a) = \sum\limits_{t=1}^{n} t + \sum\limits_{t=1}^{n} a$$

$$\sum\limits_{t=1}^{n} a = (a)(n)$$

$$\sum\limits_{t=1}^{n} t = \frac{(n)(n+1)}{2}$$

$$\sum_{t=1}^{n} t^2 = \frac{(n)(n+1)(2n+1)}{6}$$

$$\sum_{t=1}^{n} t^3 = \left(\frac{n(n+1)}{2}\right)^2$$

Example 5:

Find the sum.

$$\sum_{t=1}^{7} \left(t^2 - 8\right) =$$

$$\sum_{t=1}^{7} t^2 - \sum_{t=1}^{7} 8 = \qquad \text{We apply the rule } \sum_{t=1}^{n} (t+a) = \sum_{t=1}^{n} t + \sum_{t=1}^{n} a \,.$$

$$\left(\frac{7(8)(2 \cdot 7 + 1)}{6}\right) - (7)(8) = \qquad \text{We apply the rules } \sum_{t=1}^{n} t^2 = \frac{(n)(n+1)(2n+1)}{6} \text{ and}$$

$$\sum_{t=1}^{n} a = (a)(n).$$

$$140 - 56 = 84$$

Example 6:

Find the sum.

$$\sum_{t=1}^{5} \left(2t^3 + 5t^2 + 5t - 7\right) =$$

$$\sum_{t=1}^{5} 2t^3 + \sum_{t=1}^{5} 5t^2 + \sum_{t=1}^{5} 5t - \sum_{t=1}^{5} 7 = \qquad \text{We apply the rule } \sum_{t=1}^{n} (t+a) = \sum_{t=1}^{n} t + \sum_{t=1}^{n} a \,.$$

$$2\sum_{t=1}^{5} t^3 + 5\sum_{t=1}^{5} t^2 + 5\sum_{t=1}^{5} t - \sum_{t=1}^{5} 7 = \qquad \text{We apply the rule } \sum_{t=1}^{n} at = a\sum_{t=1}^{n} t \,.$$

$$2\left(\frac{5(5+1)}{2}\right)^2 + 5\left(\frac{5(5+1)(2 \cdot 5 + 1)}{6}\right) + 5\left(\frac{5(5+1)}{2}\right) - (5)(7) = \qquad \text{We apply the rules.}$$

$$2(225) + 5(55) + 5(15) - 35 =$$

$$450 + 275 + 75 - 35 = 765$$

» We have the sum of infinite geometric series.

$$Sum = \frac{a_1}{1-r} \qquad \text{where } |r| < 1$$

a_1 = first term

r = ratio (ratio must meet the requirement $|r| < 1$).

Example 7:
Find the sum of infinite geometric series.

$$5, \frac{15}{8}, \frac{45}{64}, \frac{135}{512}, \dots \qquad \text{The ratio} = \frac{3}{8}, \text{which meets the requirement } |r| < 1.$$

$$Sum = \frac{a_1}{1-r}$$

a_1 = first term = 5

r = ratio = $\frac{3}{8}$

$$Sum = \frac{5}{1-\frac{3}{8}} = \frac{5}{\left(\frac{5}{8}\right)} = 8$$

Chapter 7   Sequence and series

1. Find the 12th term.
 	
    4, 8, 12, 16,…
	2. Find the 8th term.
	
	    5, 10, 20, 40,…

3. Find the 38th term.
 	
    7, 10, 13, 16,…
	4. Find the 52nd term.
	
	    1, 9, 17, 25,…

5. Find the 14th term.
 	
    2, 4, 8, 16,…
	6. Find the 74th term.
	
	    1, 5, 9, 13,…

7. Find the 7th term.
 	
    3, 9, 27, 81,…
	8. Find the 9th term.
	
	    1, 4, 16, 64,…

9. Find the 15th term.
 	
    5, 12, 19, 26,…
	10. Find the 10th term.
	
	    1, 5, 25, 125,…

Chapter 7   Sequence and series

11. Find the sum of first 20 terms. 2, 6, 10, 14,…	12. Find the sum of first 42 terms. 5, 12, 19, 26,…
13. Find the sum of first 22 terms. 1, 2, 4, 8,…	14. Find the sum of first 15 terms. 5, 15, 45, 135,…
15. Find the sum of first 42 terms. 7, 12, 17, 22,…	16. Find the sum of first 11 terms. 10, 20, 40, 80,…
17. Find the sum of first 14 terms. 7, 28, 112, 448,…	18. Find the sum of first 72 terms. 18, 39, 60, 81,…
19. $\displaystyle\sum_{h=1}^{58} 21 =$	20. $\displaystyle\sum_{h=1}^{72} 2h =$

Chapter 7   Sequence and series

21. $$\sum_{h=1}^{24}\left(h^2-4h+2\right)=$$	22. $$\sum_{h=1}^{32}\left(h^2-5\right)=$$
23. $$\sum_{h=1}^{45}\left(7h^3\right)=$$	24. $$\sum_{h=1}^{24}\left(2h^3-h^2+2h+1\right)=$$
25. $$\sum_{h=1}^{27}\left(5h^3+2h^2-3h+4\right)=$$	26. Find the sum of infinite geometric series $$7+\frac{7}{3}+\frac{7}{9}+\frac{7}{27}+\cdots$$
27. Find the sum of infinite geometric series $$10-5+\frac{5}{2}-\frac{5}{4}+\cdots$$	28. Find the sum of infinite geometric series $$3+\frac{3}{5}+\frac{3}{25}+\frac{3}{125}+\cdots$$
29. Find the sum of infinite geometric series $$11+\frac{44}{5}+\frac{176}{25}+\frac{704}{125}+\cdots$$	30. Find the sum of infinite geometric series $$7-1+\frac{1}{7}-\frac{1}{49}+\cdots$$

Chapter 7 Sequence and series
Answer key

1. $a_n = a_1 + (n-1)d$
   $a_{12} = 4 + (12-1)(4)$
   $a_{12} = 48$

2. $a_n = a_1 r^{n-1}$
   $a_8 = 5(2)^{8-1}$
   $a_8 = 640$

3. $a_n = a_1 + (n-1)d$
   $a_{38} = 7 + (38-1)(3)$
   $a_{38} = 118$

4. $a_n = a_1 + (n-1)d$
   $a_{52} = 1 + (52-1)(8)$
   $a_{52} = 409$

5. $a_n = a_1 r^{n-1}$
   $a_{14} = 2(2)^{14-1}$
   $a_{14} = 16,384$

6. $a_n = a_1 + (n-1)d$
   $a_{74} = 1 + (74-1)(4)$
   $a_{74} = 293$

7. $a_n = a_1 r^{n-1}$
   $a_7 = 3(3)^{7-1}$
   $a_7 = 2,187$

8. $a_n = a_1 r^{n-1}$
   $a_9 = 1(4)^{9-1}$
   $a_9 = 65,536$

9. $a_n = a_1 + (n-1)d$
   $a_{15} = 5 + (15-1)(7)$
   $a_{15} = 103$

10. $a_n = a_1 r^{n-1}$
    $a_{10} = 1(5)^{10-1}$
    $a_{10} = 1,953,125$

11. $S_n = \dfrac{n}{2}(first.term + last.term)$
    $a_{20} = 2 + (20-1)(4)$
    $a_{20} = 78$
    $S_{20} = \dfrac{20}{2}(2+78) = 800$

12. $S_n = \dfrac{n}{2}(first.term + last.term)$
    $a_{42} = 5 + (42-1)(7)$
    $a_{42} = 292$
    $S_{42} = \dfrac{42}{2}(5+292) = 6,237$

13. $S_n = a_1\left(\dfrac{1-r^n}{1-r}\right)$
    $S_{22} = 1\left(\dfrac{1-2^{22}}{1-2}\right) = 4,194,303$

14. $S_n = a_1\left(\dfrac{1-r^n}{1-r}\right)$
    $S_{15} = (5)\left(\dfrac{1-3^{15}}{1-3}\right) = 35,872,265$

15. $S_n = \dfrac{n}{2}(first.term + last.term)$
    $a_{42} = 7 + (42-1)(5)$
    $a_{42} = 212$
    $S_{42} = \dfrac{42}{2}(7+212) = 4,599$

16. $S_n = a_1\left(\dfrac{1-r^n}{1-r}\right)$

$S_{11} = 10\left(\dfrac{1-2^{11}}{1-2}\right)$

$S_{11} = 20{,}470$

17. $S_n = a_1\left(\dfrac{1-r^n}{1-r}\right)$

$S_{14} = 7\left(\dfrac{1-4^{14}}{1-4}\right)$

$S_{14} = 626{,}349{,}395$

18. $S_n = \dfrac{n}{2}\left(first.term + last.term\right)$

$S_{72} = 18 + (72-1)(21)$

$a_{72} = 1509$

$S_{72} = \dfrac{72}{2}(18 + 1{,}509) = 54{,}972$

19. $\displaystyle\sum_{h=1}^{58} 21 = (58)(21) = 1{,}218$

20. $\displaystyle\sum_{h=1}^{72}(2h) = 2\sum_{h=1}^{72} h = 2\left[\dfrac{72(73)}{2}\right] = 5{,}256$

21. $\displaystyle\sum_{h=1}^{24}\left(h^2 - 4h + 2\right) =$

$\displaystyle\sum_{h=1}^{24} h^2 - 4\sum_{h=1}^{24} h + \sum_{h=1}^{24} 2 =$

$\dfrac{24(24+1)(24\cdot 2+1)}{6} - 4\left[\dfrac{24(25)}{2}\right] + (24)(2)$

$= 4{,}900 - 1{,}200 + 48 = 3{,}748$

22. $\displaystyle\sum_{h=1}^{32}\left(h^2 - 5\right) = \sum_{h=1}^{32} h^2 - \sum_{h=1}^{32} 5 =$

$\dfrac{(32)(32+1)(32\cdot 2+1)}{6} - (32)(5) =$

$11{,}440 - 160 = 11{,}280$

23. $\displaystyle\sum_{h=1}^{45}\left(7h^3\right) = 7\sum_{h=1}^{45} h^3 =$

$7\left[\left(\dfrac{45\cdot 46}{2}\right)^2\right] = 7{,}498{,}575$

24. $\displaystyle\sum_{h=1}^{24}\left(2h^3 - h^2 + 2h + 1\right) =$

$2\displaystyle\sum_{h=1}^{24} h^3 - \sum_{h=1}^{24} h^2 + 2\sum_{h=1}^{24} h + \sum_{h=1}^{24} 1 =$

$2\left[\left(\dfrac{24\cdot 25}{2}\right)^2\right] - \dfrac{24(25)(2\cdot 24+1)}{6} + 2\left[\dfrac{24\cdot 25}{2}\right] + 24(1)$

$= 180{,}000 - 4{,}900 + 600 + 24 = 175{,}724$

25. $\displaystyle\sum_{h=1}^{27}\left(5h^3 + 2h^2 - 3h + 4\right) =$

$5\displaystyle\sum_{h=1}^{27} h^3 + 2\sum_{h=1}^{27} h^2 - 3\sum_{h=1}^{27} h + \sum_{h=1}^{27} 4 =$

$5\left[\left(\dfrac{27\cdot 28}{2}\right)^2\right] + 2\left[\dfrac{(27)(28)(27\cdot 2+1)}{6}\right] - 3\left[\dfrac{27\cdot 28}{2}\right] + (27)(4)$

$= 714{,}420 + 13{,}860 - 1{,}134 + 108 = 727{,}254$

26. $sum = \dfrac{a_1}{1-r} = \dfrac{7}{1-\dfrac{1}{3}} = \dfrac{21}{2}$

27. $sum = \dfrac{a_1}{1-r} = \dfrac{10}{1-\left(\dfrac{-1}{2}\right)} = \dfrac{20}{3}$

28. $sum = \dfrac{a_1}{1-r} = \dfrac{3}{1-\left(\dfrac{1}{5}\right)} = \dfrac{15}{4}$

29. $sum = \dfrac{a_1}{1-r} = \dfrac{11}{1-\left(\dfrac{4}{5}\right)} = 55$

30. $sum = \dfrac{a_1}{1-r} = \dfrac{7}{1-\left(\dfrac{-1}{7}\right)} = \dfrac{49}{8}$

Chapter 8   Permutation, combination, and probability

▶ When we talk about arrangements, we can study the permutation and combination. Permutation is arrangements where order matters, whereas order does not matter in combination.

Permutation has a formula.

$$_nP_r = \frac{n!}{(n-r)} \qquad \text{where } n \geq r$$

Combination has a formula.

$$_nC_r = \frac{n!}{r!(n-r)}$$

Example 1:
Evaluate the permutation.

$$_8P_3 =$$

$$_nP_r = \frac{n!}{(n-r)} \qquad \text{We apply the permutation formula.}$$

$$_8P_3 = \frac{8!}{(8-3)} = \frac{8!}{5!} \qquad n \text{ is always the larger or at least the equal value of } r.$$

$$\frac{8!}{5!} = \frac{8\cdot7\cdot6\cdot5\cdot4\cdot3\cdot2\cdot1}{5\cdot4\cdot3\cdot2\cdot1} \qquad \text{! is factorial. This means we start from the number}$$

and multiply one-by-one down to number 1.

$$= \frac{8\cdot7\cdot6\cdot5\cdot4\cdot3\cdot2\cdot1}{5\cdot4\cdot3\cdot2\cdot1} \qquad \text{Common factors in the numerator and in the}$$

denominator cancel out.

$$= 8\cdot7\cdot6 = 336$$

Example 2:
Evaluate the combination.

$$_7C_5 =$$

$$_nC_r = \frac{n!}{r!(n-r)} \qquad \text{We apply the combination formula.}$$

$$_7C_5 = \frac{7!}{5!(7-5)} = \frac{7!}{5!\cdot 2!}$$

For factorial !, we start from the number and multiply one-by-one down to number 1.

$$\frac{7!}{5!\cdot 2!} = \frac{7\cdot 6\cdot 5\cdot 4\cdot 3\cdot 2\cdot 1}{(5\cdot 4\cdot 3\cdot 2\cdot 1)(2\cdot 1)}$$

$$= \frac{7\cdot 6^3\cdot 5\cdot 4\cdot 3\cdot 2\cdot 1}{(5\cdot 4\cdot 3\cdot 2\cdot 1)(2\cdot 1)}$$

Common factors in the numerator and in the denominator cancel.

$$= 7\cdot 3 = 21$$

Example 3:
Evaluate the expression.

$$\left(_4P_3\right)\left(_5C_2\right)=$$

We find combination and permutation, and then, we multiply the results.

$$_nP_r = \frac{n!}{(n-r)}$$

$$_4P_3 = \frac{4!}{(4-3)} = \frac{4!}{1!} = \frac{4\cdot 3\cdot 2\cdot 1}{1} = 24$$

$$_nC_r = \frac{n!}{r!(n-r)}$$

$$_5C_2 = \frac{5!}{2!(5-2)} = \frac{5!}{2!\cdot 3!} = \frac{5\cdot 4^2\cdot 3\cdot 2\cdot 1}{(2\cdot 1)(3\cdot 2\cdot 1)} = 5\cdot 2 = 10$$

$$_4P_3 = 24 \text{ and } _5C_2 = 10$$

$$\left(_4P_3\right)\left(_5C_2\right) = (24)(10) = 240$$

» Now that we have learned the permutation and combination formulas, we will study how to apply them.

Example 4:
A telephone company is assigning 7-digit telephone numbers.
The first and second digits must be odd numbers. The
third and fourth digits must be even numbers. If numbers

cannot be repeated, how many different numbers can the telephone company assign?

— — — — — — —

We create seven spaces, representing the digits in order. We have ten digits from 0 to 9. We have five even digits: 0, 2, 4, 6, and 8. We also have five odd numbers: 1, 3, 5, 7, and 9.

5 — — — — — —

The first two digits must be odd. For the first digit, any of the five odd numbers (1, 3, 5, 7, and 9) may be selected. We write 5 in the first digit because we have five possible choices.

5 4 — — — — —

For the second digit, any of the remaining four odd numbers may be selected. Remember that the numbers cannot repeat in this question. We write 4 in the second digit because, once we pick one odd number in the first digit, we have four remaining odd numbers.

5 4 5 — — — —

The third and fourth digits are even numbers. Any one of the five even numbers (0, 2, 4, 6, and 8) may be selected for the third digit. We write 5 in the third digit because we have five possibilities.

5 4 5 4 — — —

Any of the remaining four even numbers may be selected for the fourth digit. The numbers cannot repeat. Because we have selected one even number in the third digit, we have four remaining even numbers. We write 4 in the fourth digit.

5 4 5 4 6 — —

Fifth, sixth, and seventh digits can be either odd or even. Because we have selected four numbers in the previous four digits, we have six remaining numbers. For fifth digit, any of the six remaining numbers may be selected. We write 6 in the fifth digit.

<u>5</u> <u>4</u> <u>5</u> <u>4</u> <u>6</u> <u>5</u> __

We have five numbers remaining because we selected five numbers in the previous five digits. For the sixth digit, any of the remaining five numbers may be selected. We write 5 in the sixth digit.

<u>5</u> <u>4</u> <u>5</u> <u>4</u> <u>6</u> <u>5</u> <u>4</u>

For the last digit, we have four remaining numbers because six other numbers were selected in the previous six digits. We write 4 in the seventh digit.

» Probability is a likelihood of an event happening. We can apply the probability formula.

$$p(\text{event}) = \frac{\text{favorable outcome}}{\text{total possible outcome}}$$

Based on this formula, we can derive another equation. We move the denominator "total possible outcome" to the other side of the equation. The p(event) times the "total possible outcome" is equal to the "favorable outcome":

$$p(\text{event}) \times \text{total possible outcome} = \text{favorable outcome}$$

Example 5:

The probability of picking a green marble is $\frac{2}{15}$.

If there are 300 marbles in the jar, how many green marbles are there?

$$p(\text{event}) \times \text{total possible outcome} = \text{favorable outcome}$$

$$p(\text{event}) = \frac{2}{15}$$

total possible outcome $= 300$

$$\left(\frac{2}{15}\right)(300) = \text{favorable outcome} = 40 \qquad \text{We have 40 green marbles.}$$

» We can find the probability of an event not happening by subtracting probability of even happening from 1.

$$p(\text{event not happening}) = 1 - p(\text{event}) = 1 - \frac{\text{favorable outcome}}{\text{total possible outcome}}$$

Example 6:
Out of 28 flash cards, 11 cards have letter A.
What is the probability of not picking a card
with letter A?

$$p(\text{event not happening}) = 1 - p(\text{event}) = 1 - \frac{\text{favorable outcome}}{\text{total possible outcome}}$$

$$p(\text{letter A}) = \frac{11}{28}$$

We first identify the probability of the event happening.

$$p(\text{not letter A}) = 1 - \frac{11}{28} = \frac{17}{28}$$

Then, we subtract the probability of an event happening from 1.

$$p(\text{not letter A}) = \frac{17}{28}$$

» Probability is a likelihood of an event happening. We can apply the probability formula.

$$p(\text{event}) = \frac{\text{favorable outcome}}{\text{total possible outcome}}$$

Example 7:
A number is randomly picked from the number
set {2, 7, 8, 10, 14, 23, 25, 27, 28}. What is the
probability of picking an even number?

$$p(\text{event}) = \frac{\text{favorable outcome}}{\text{total possible outcome}}$$

favorable outcome = 5

We have five even numbers (2, 8, 10, 14, 28).

total possible outcome = 9

$$p(\text{even number}) = \frac{5}{9}$$

Chapter 8   Permutation, combination, and probability

1. Evaluate the permutation. $_8P_5 =$	2. Evaluate the combination. $_{11}C_8 =$
3. Evaluate the expression. $$\frac{\left(_5P_3\right)\left(_3C_2\right)}{_4P_1} =$$	4. Evaluate the expression. $$\left(_3P_1\right)\left(_7P_4\right)\left(_4C_3\right) =$$
5. Evaluate the expression. $$\frac{\left(_7P_4\right)}{\left(_4C_3\right)\left(_3P_2\right)} =$$	6. A company is creating ID numbers for its employees. The number has four digits and it does not contain 0, 7, and 8. If numbers cannot be repeated, how many possible ID numbers can be created?
7. A telephone company is assigning 7-digit telephone numbers. The first digit must be an odd number and the second digit must be an even number. If numbers cannot be repeated, how many different numbers can the telephone company assign?	8. A school is assigning 4-digit locker numbers. The first two digits must be even numbers and the last two digits must be odd numbers. If numbers cannot be repeated, how many different numbers can the school assign?
9. A school is assigning 6-spaced student ID codes. First two spaces are for alphabets and the last four spaces are for numbers. If alphabets and numbers can be repeated, how many different student ID codes can the school create?	10. A company is assigning 5-spaced employee ID codes. First three spaces are for alphabets and the last two spaces are for numbers. If alphabets and numbers cannot be repeated, how many different employee ID codes can the company create?

Chapter 8   Permutation, combination, and probability

11. The probability of picking a green marble is $\frac{2}{7}$. If there are 24 green marbles in the jar, how many marbles are there?	12. The probability of picking a blue marble is $\frac{5}{7}$. If there are 15 blue marbles in the jar, how many marbles are there?
13. The probability of picking a blue marble is $\frac{11}{17}$. If there are 44 blue marbles in the jar, how many marbles are there?	14. The probability of picking a beige marble is $\frac{7}{8}$. If there are 56 marbles in the jar, how many are beige marbles?
15. The probability of picking a blue marble is $\frac{3}{5}$. If there are 55 marbles in the jar, how many blue marbles are there?	16. The probability of picking a blue marble is $\frac{2}{7}$. If there are 28 marbles in the jar, how many blue marbles are there?
17. Out of 14 flash cards, 6 cards have letter A. What is the probability of not picking a card with letter A?	18. Out of 15 flash cards, 7 cards have letter E. What is the probability of not picking a card with letter E?
19. Out of 42 flash cards, 23 cards have letter D. What is the probability of not picking a card with letter D?	20. Out of 102 flash cards, 58 cards have letter B. What is the probability of not picking a card with letter B?

Chapter 8   Permutation, combination, and probability

21. A number is randomly picked from the number set $\{1, 5, 11, 12, 14, 17, 18, 41\}$ What is the probability of picking an odd number?	22. A number is randomly picked from the number set $\{5, 17, 22, 31, 34, 172, 178\}$ What is the probability of picking a two-digit number?
23. A number is randomly picked from the number set $\{-2, -4, 7, -11, 4, 24, -1, -15\}$ What is the probability of picking a negative number?	24. A number is randomly picked from the number set $\{25, -3, 0, -17, 4, 1, 8\}$ What is the probability of picking a positive number?
25. A number is randomly picked from the number set $\{-3, 0, 4, 27, -24, -37, -1\}$ What is the probability of picking a number that is both odd and negative?	26. A number is randomly picked from the number set $\{-17, \frac{2}{3}, 0, -\frac{1}{4}, \frac{1}{5}, \frac{3}{5}, 12\}$ What is the probability of picking a number that is both fraction and positive?
27. A number is randomly picked from the number set $\{7, 11, 12, -142, 72, 1502\}$ What is the probability of picking a number that is both odd and two-digit?	28. A number is randomly picked from the number set $\{-5, 0, 4, 2, -11, 18\}$ What is the probability of picking a number that is both odd and positive?
29. A number is randomly picked from the number set $\{3, 8, -4, 7, 14, -12, 11, 5, 102\}$ What is the probability of picking a number that is both negative and even?	30. A number is randomly picked from the number set $\{-2, -4, 0, \frac{3}{8}, 2, \frac{4}{7}, -\frac{11}{12}, 8, -\frac{1}{2}, \frac{3}{7}\}$ What is the probability of picking a positive integer?

Chapter 8 Permutation, combination, and probability
Answer key

1. $_8P_5 = \dfrac{8!}{(8-5)} = \dfrac{8!}{3!} =$

$\dfrac{8 \cdot 7 \cdot 6 \cdot 5 \cdot 4 \cdot 3 \cdot 2 \cdot 1}{3 \cdot 2 \cdot 1} = 6{,}720$

2. $_{11}C_8 = \dfrac{11!}{8!(11-8)} = \dfrac{11!}{8! \cdot 3!} =$

$\dfrac{11 \cdot^5 10 \cdot^3 9 \cdot 8 \cdot 7 \cdots 1}{(8 \cdot 7 \cdots 1)(3 \cdot 2 \cdot 1)} = 11 \cdot 5 \cdot 3 = 165$

3. $\dfrac{(_5P_3)(_3C_2)}{_4P_1} = \dfrac{\left(\dfrac{5!}{(5-3)}\right)\left(\dfrac{3!}{2!(3-2)}\right)}{\left(\dfrac{4!}{(4-1)}\right)} =$

$\dfrac{\left(\dfrac{5!}{2!}\right)\left(\dfrac{3!}{2! \cdot 1!}\right)}{\left(\dfrac{4!}{3!}\right)} = \dfrac{\left(\dfrac{5 \cdot 4 \cdot 3 \cdot 2 \cdot 1}{2 \cdot 1}\right)\left(\dfrac{3 \cdot 2 \cdot 1}{2 \cdot 1}\right)}{\left(\dfrac{4 \cdot 3 \cdot 2 \cdot 1}{3 \cdot 2 \cdot 1}\right)} =$

$= \dfrac{(5 \cdot 4 \cdot 3)(3)}{4} = 5 \cdot 3 \cdot 3 = 45$

4. $(_3P_1)(_7P_4)(_4C_3) =$

$\left(\dfrac{3!}{(3-1)}\right)\left(\dfrac{7!}{(7-4)}\right)\left(\dfrac{4!}{3!(4-3)}\right) =$

$\left(\dfrac{3!}{2!}\right)\left(\dfrac{7!}{3!}\right)\left(\dfrac{4!}{3! \cdot 1!}\right) = 3 \cdot 7 \cdot 6 \cdot 5 \cdot 4 \cdot 4$

$= 10{,}080$

5. $\dfrac{(_7P_4)}{(_4C_3)(_3P_2)} = \dfrac{\dfrac{7!}{(7-4)}}{\left(\dfrac{4!}{3!(4-3)}\right)\left(\dfrac{3!}{(3-2)}\right)} =$

$\dfrac{\dfrac{7!}{3!}}{\left(\dfrac{4!}{3! \cdot 1!}\right)\left(\dfrac{3!}{1!}\right)} = 7 \cdot 5 = 35$

6. $\underline{7} \cdot \underline{6} \cdot \underline{5} \cdot \underline{4} = 840$

7. $\underline{5} \cdot \underline{5} \cdot \underline{8} \cdot \underline{7} \cdot \underline{6} \cdot \underline{5} \cdot \underline{4} = 168{,}000$

8. $\underline{5} \cdot \underline{4} \cdot \underline{5} \cdot \underline{4} = 400$

9. $\underline{26} \cdot \underline{26} \cdot \underline{10} \cdot \underline{10} \cdot \underline{10} \cdot \underline{10} = 6{,}760{,}000$

10. $\underline{26} \cdot \underline{25} \cdot \underline{24} \cdot \underline{10} \cdot \underline{9} = 1{,}404{,}000$

11. $\dfrac{2}{7}X = 24$

$X = 24\left(\dfrac{7}{2}\right) = 84$

84 marbles

12. $\dfrac{5}{7}X = 15$

$X = 15\left(\dfrac{7}{5}\right) = 21$

21 marbles

13. $\dfrac{11}{17}X = 44$

$X = 44\left(\dfrac{17}{11}\right) = 68$

68 marbles

14. $\dfrac{7}{8}(56) = 49$

49 beige marbles

15. $\dfrac{3}{5}(55) = 33$

    33 blue marbles

16. $\left(\dfrac{2}{7}\right)(28) = 8$

    8 blue marbles

17. $p(\textit{letter A}) = \dfrac{6}{14}$

    $p(\textit{Not letter A}) = 1 - \dfrac{6}{14} = \dfrac{4}{7}$

18. $p(\textit{letter E}) = \dfrac{7}{15}$

    $p(\textit{Not letter E}) = 1 - \dfrac{7}{15} = \dfrac{8}{15}$

19. $p(\textit{letter D}) = \dfrac{23}{42}$

    $p(\textit{Not letter D}) = 1 - \dfrac{23}{42} = \dfrac{19}{42}$

20. $p(\textit{letter B}) = \dfrac{58}{102}$

    $p(\textit{Not letter B}) = 1 - \dfrac{58}{102} = \dfrac{22}{51}$

21. $p(\textit{odd number}) = \dfrac{5}{8}$

22. $p(\textit{two-digit number}) = \dfrac{4}{7}$

23. $p(\textit{negative number}) = \dfrac{5}{8}$

24. Note that 0 is neither negative nor positive.

    $p(\textit{positive number}) = \dfrac{4}{7}$

25. $p(\textit{odd and negative}) = \dfrac{3}{7}$

26. $p(\textit{fraction and positive}) = \dfrac{3}{7}$

27. $p(\textit{odd and two-digit}) = \dfrac{1}{6}$

28. $p(\textit{odd and positive}) = \dfrac{0}{6} = 0$

29. $p(\textit{negative and even}) = \dfrac{2}{9}$

30. An integer is neither fraction nor decimal.

    $p(\textit{positive integer}) = \dfrac{2}{10} = \dfrac{1}{5}$

Chapter 9   Imaginary numbers

▶ The standard form of imaginary number is $a + bi$ where a is the real part and b is the imaginary part. When we add or subtract imaginary numbers, we combine the real parts and combine the imaginary parts.

Example 1:

$$(21 + 8i) + (7 + 12i) =$$ We are adding imaginary numbers.

$$[21 + 7] + [8i + 12i] =$$ We combine the real parts and combine the imaginary parts.

$$28 + 20i$$

Example 2:

$$(1 + 8i) - (2 + 7i) =$$ We are subtracting two imaginary numbers.

$$1 + 8i - 2 - 7i =$$ We first distribute the negative.

$$(1 - 2) + (8i - 7i) =$$ We combine the real parts and combine the imaginary parts.

$$-1 + i$$

Example 3:

$$(7 - 8i) - (21 + 22i) + (15 + 7i) =$$

$$7 - 8i - 21 - 22i + 15 + 7i =$$ We distribute.

$$(7 - 21 + 15) + (-8i - 22i + 7i) =$$ We combine the real parts and combine the imaginary parts.

$$1 - 23i$$

» We can multiply imaginary numbers. We apply distribution method or FOIL method when we multiply imaginary numbers. When multiplying, we will see i^2, which is equal to -1.

$$i^2 = -1$$

Imaginary value repeats every fourth power, as shown below.

$$i = \sqrt{-1} \qquad\qquad i^5 = \sqrt{-1} \qquad\qquad i^9 = \sqrt{-1}$$
$$i^2 = -1 \qquad\qquad i^6 = -1 \qquad\qquad i^{10} = -1$$
$$i^3 = -i \qquad\qquad i^7 = -i \qquad\qquad i^{11} = -i$$
$$i^4 = 1 \qquad\qquad i^8 = 1 \qquad\qquad i^{12} = 1$$

Example 4:

$$(20 - 2i)(5 + 2i) =$$

$$(20)(5) + (20)(2i) + (-2i)(5) + (-2i)(2i) = \qquad \text{We apply the FOIL method. We}$$
$$\text{multiply first, outer, inner, and last.}$$

$$100 + 40i - 10i - 4i^2 =$$

$$100 + 40i - 10i - 4(-1) = \qquad \text{We apply } i^2 = -1.$$

$$100 + 40i - 10i + 4 = \qquad \text{We combine the like terms.}$$

$$104 + 30i$$

» We will study about rationalizing complex number fraction. We do not want to have a complex number in the denominator. We rationalize by multiplying denominator's conjugate in the numerator and in the denominator.

Conjugate of $A + B$ is $A - B$.
Conjugate of $A - B$ is $A + B$.
We switch the signs to find the conjugate.

Example 5:
Multiply by conjugates.

$$\frac{12}{15 + 8i} \qquad\qquad \text{We rationalize.}$$

$$\frac{12}{15 + 8i}\left(\frac{15 - 8i}{15 - 8i}\right) = \qquad \text{We multiply by conjugate to remove } i \text{ in the denominator.}$$

$$\frac{12(15 - 8i)}{(15 + 8i)(15 - 8i)} = \frac{180 - 96i}{225 - 120i + 120i - 64i^2} =$$

$$\frac{180 - 96i}{225 - 64i^2} = \frac{180 - 96i}{225 - 64(-1)} =$$

$$\frac{180 - 96i}{289} =$$

$$\frac{180}{289} - \frac{96}{289}i \qquad \text{We simplify the expression to the standard form } a + bi.$$

Example 6:
Multiply by conjugates.

$$\frac{3 + 15i}{12 - 8i} \qquad \text{We rationalize.}$$

$$\left(\frac{3 + 15i}{12 - 8i}\right)\left(\frac{12 + 8i}{12 + 8i}\right) = \qquad \text{We multiply by conjugates.}$$

$$\frac{(3 + 15i)(12 + 8i)}{(12 - 8i)(12 + 8i)} = \frac{36 + 24i + 180i + 120i^2}{144 + 96i - 96i - 64i^2}$$

$$\frac{36 + 204i + 120(-1)}{144 - 64(-1)} = \frac{36 + 204i - 120}{144 + 64} =$$

$$\frac{-84 + 204i}{208} = \frac{-84}{208} + \frac{204}{208}i =$$

$$-\frac{21}{52} + \frac{51}{52}i \qquad \text{We simplify the expression to the standard form } a + bi.$$

» When we solved quadratic equation, we had three possibilities: 1) Two solutions, 2). One solution, or 3). No solution.

For quadratic equations with no solutions, we can express the complex numbers (imaginary numbers) as solutions. To do this we apply the quadratic formula.

$$\frac{-b \pm \sqrt{b^2 - 4ac}}{2a} \quad \text{and the imaginary value } \sqrt{-1} = i$$

Example 7:
Find the solutions in terms of imaginary numbers.

$$-2X^2 - 3X - 10 = 0 \qquad \text{We apply the quadratic formula.}$$

$$\frac{-b \pm \sqrt{b^2 - 4ac}}{2a}$$

$aX^2 + bX + c = 0$ — We identify the values of a, b, and c.

$-2X^2 - 3X - 10 = 0$

$a = -2$, $b = -3$, $c = -10$

$$X = \frac{-(-3) \pm \sqrt{(-3)^2 - 4(-2)(-10)}}{2(-2)}$$ — We substitute the corresponding values.

$$= \frac{3 \pm \sqrt{-71}}{-4}$$ — Then, we evaluate.

$$= \frac{3 \pm \sqrt{71}\sqrt{-1}}{-4} = \frac{3 \pm i\sqrt{71}}{-4} = -\frac{3}{4} \pm \frac{i\sqrt{71}}{4}$$

$$= -\frac{3}{4} - \frac{i\sqrt{71}}{4} \text{ and } -\frac{3}{4} + \frac{i\sqrt{71}}{4}$$ — We express in standard form $a + bi$.

Chapter 9   Imaginary numbers

1. $(12 - 2i) + (5 + 4i) =$	2. $(5 - i) - (7 - 15i) =$
3. $(2 + 23i) + (5 - 20i) =$	4. $(21 + 5i) + (42 + 25i) =$
5. $(57 - 2i) - (48 + 12i) =$	6. $(10 - 10i) - (12 + 4i) =$
7. $(27 - i) - (1 + 2i) =$	8. $(-15 + 2i) - (14 - i) =$
9. $(2 + i)(1 - 12i) =$	10. $(5 - 4i)(3 + 2i) =$

Chapter 9   Imaginary numbers

11. $(12-i)(14+2i)=$	12. $(5+10i)(4-11i)=$
13. $(4+21i)(1-i)=$	14. $(12+20i)(1+2i)=$
15. $(8-3i)(5+i)=$	16. Multiply by conjugates. $\dfrac{2}{1+5i}$
17. Multiply by conjugates. $\dfrac{5}{1-10i}$	18. Multiply by conjugates. $\dfrac{4}{2+3i}$
19. Multiply by conjugates. $\dfrac{1-i}{2+5i}$	20. Multiply by conjugates. $\dfrac{14+2i}{5-4i}$

Chapter 9   Imaginary numbers

21. Multiply by conjugates. $\dfrac{5-i}{11+10i}$	22. Multiply by conjugates. $\dfrac{1+i}{2-11i}$
23. Multiply by conjugates. $\dfrac{3-4i}{1+i}$	24. Find the solutions in terms of imaginary numbers. $X^2+7X+13=0$
25. Find the solutions in terms of imaginary numbers. $X^2-2X+5=0$	26. Find the solutions in terms of imaginary numbers. $5X^2-X+1=0$
27. Find the solutions in terms of imaginary numbers. $3X^2+5X+10=0$	28. Find the solutions in terms of imaginary numbers. $X^2+5X+11=0$
29. Find the solutions in terms of imaginary numbers. $7X^2-X+2=0$	30. Find the solutions in terms of imaginary numbers. $6X^2-X+5=0$

Chapter 9 Imaginary numbers
Answer key

1. $(12-2i)+(5+4i)=$
$12-2i+5+4i=17+2i$

2. $(5-i)-(7-15i)=$
$5-i-7+15i=-2+14i$

3. $(2+23i)+(5-20i)=$
$2+23i+5-20i=7+3i$

4. $(21+5i)+(42+25i)=$
$21+5i+42+25i=63+30i$

5. $(57-2i)-(48+12i)=$
$57-2i-48-12i=9-14i$

6. $(10-10i)-(12+4i)=$
$10-10i-12-4i=-2-14i$

7. $(27-i)-(1+2i)=$
$27-i-1-2i=26-3i$

8. $(-15+2i)-(14-i)=$
$-15+2i-14+i=-29+3i$

9. $(2+i)(1-12i)=$
$2-24i+i-12i^2=$
$2-23i-12(-1)=$
$14-23i$

10. $(5-4i)(3+2i)=$
$15+10i-12i-8i^2=$
$15-2i-8(-1)=$
$23-2i$

11. $(12-i)(14+2i)=$
$168+24i-14i-2i^2=$
$168+10i-2(-1)=$
$170+10i$

12. $(5+10i)(4-11i)=$
$20-55i+40i-110i^2=$
$20-15i-110(-1)=130-15i$

13. $(4+21i)(1-i)=$
$4-4i+21i-21i^2=$
$4+17i-21(-1)=25+17i$

14. $(12+20i)(1+2i)=$
$12+24i+20i+40i^2=$
$12+44i+40(-1)=-28+44i$

15. $(8-3i)(5+i)=$
$40+8i-15i-3i^2=$
$40-7i-3(-1)=43-7i$

16. $\dfrac{2}{1+5i}\left(\dfrac{1-5i}{1-5i}\right)=\dfrac{2-10i}{1-25i^2}=$
$\dfrac{2-10i}{1-25(-1)}=\dfrac{2-10i}{26}=\dfrac{1}{13}-\dfrac{5}{13}i$

17. $\dfrac{5}{1-10i}\left(\dfrac{1+10i}{1+10i}\right)=\dfrac{5+50i}{1-100i^2}=$
$\dfrac{5+50i}{1-100(-1)}=\dfrac{5+50i}{101}=\dfrac{5}{101}+\dfrac{50}{101}i$

18. $\dfrac{4}{2+3i}\left(\dfrac{2-3i}{2-3i}\right)=$
$\dfrac{8-12i}{4-9i^2}=\dfrac{8-12i}{4-9(-1)}=\dfrac{8-12i}{13}$
$=\dfrac{8}{13}-\dfrac{12}{13}i$

19. $\dfrac{1-i}{2+5i}\left(\dfrac{2-5i}{2-5i}\right)=$
$\dfrac{2-5i-2i+5i^2}{4-25i^2}=\dfrac{2-7i-5}{4-25(-1)}$
$=\dfrac{-3-7i}{29}=\dfrac{-3}{29}-\dfrac{7}{29}i$

20.

$$\frac{14+2i}{5-4i}\left(\frac{5+4i}{5+4i}\right)=\frac{70+56i+10i+8i^2}{25-16i^2}=$$

$$\frac{70+66i+8(-1)}{25-16(-1)}=\frac{62+66i}{41}=\frac{62}{41}+\frac{66}{41}i$$

21. $\dfrac{5-i}{11+10i}\left(\dfrac{11-10i}{11-10i}\right)=$

$$\frac{55-50i-11i+10i^2}{121-100i^2}=\frac{55-61i+10(-1)}{121-100(-1)}$$

$$=\frac{45-61i}{221}=\frac{45}{221}-\frac{61}{221}i$$

22. $\dfrac{1+i}{2-11i}\left(\dfrac{2+11i}{2+11i}\right)=\dfrac{2+11i+2i+11i^2}{4-121i^2}$

$$=\frac{2+13i+11(-1)}{4-121(-1)}=\frac{-9+13i}{125}=\frac{-9}{125}+\frac{13}{125}i$$

23. $\dfrac{3-4i}{1+i}\left(\dfrac{1-i}{1-i}\right)=\dfrac{3-3i-4i+4i^2}{1-i^2}$

$$=\frac{3-7i+4(-1)}{1-(-1)}=\frac{-1-7i}{2}=\frac{-1}{2}-\frac{7}{2}i$$

24. $\dfrac{-b\pm\sqrt{b^2-4ac}}{2a}$

$$\frac{-7\pm\sqrt{7^2-4(1)(13)}}{2(1)}=\frac{-7\pm\sqrt{-3}}{2}=\frac{-7\pm i\sqrt{3}}{2}$$

$$=\frac{-7}{2}-\frac{\sqrt{3}}{2}i \ \ \text{and} \ \ \frac{-7}{2}+\frac{\sqrt{3}}{2}i$$

25. $\dfrac{-b\pm\sqrt{b^2-4ac}}{2a}$

$$\frac{-(-2)\pm\sqrt{(-2)^2-4(1)(5)}}{2(1)}=\frac{2\pm\sqrt{-16}}{2}$$

$$=\frac{2\pm4i}{2}=1\pm2i=$$

$$1-2i \ \ \text{and} \ \ 1+2i$$

26. $\dfrac{-b\pm\sqrt{b^2-4ac}}{2a}$

$$\frac{-(-1)\pm\sqrt{(-1)^2-4(5)(1)}}{2(5)}=\frac{1\pm\sqrt{-19}}{10}=\frac{1\pm i\sqrt{19}}{10}$$

$$=\frac{1}{10}-\frac{\sqrt{19}}{10}i \ \ \text{and} \ \ \frac{1}{10}+\frac{\sqrt{19}}{10}i$$

27. $\dfrac{-b\pm\sqrt{b^2-4ac}}{2a}$

$$\frac{-5\pm\sqrt{(5)^2-4(3)(10)}}{2(3)}=\frac{-5\pm\sqrt{-95}}{6}=\frac{-5\pm i\sqrt{95}}{6}$$

$$=\frac{-5}{6}-\frac{\sqrt{95}}{6}i \ \ \text{and} \ \ \frac{-5}{6}+\frac{\sqrt{95}}{6}i$$

28. $\dfrac{-b\pm\sqrt{b^2-4ac}}{2a}$

$$\frac{-5\pm\sqrt{(5)^2-4(1)(11)}}{2(1)}=\frac{-5\pm\sqrt{-19}}{2}=\frac{-5\pm i\sqrt{19}}{2}$$

$$=\frac{-5}{2}-\frac{\sqrt{19}}{2}i \ \ \text{and} \ \ \frac{-5}{2}+\frac{\sqrt{19}}{2}i$$

29. $\dfrac{-b\pm\sqrt{b^2-4ac}}{2a}$

$$\frac{-(-1)\pm\sqrt{(-1)^2-4(7)(2)}}{2(7)}=\frac{1\pm\sqrt{-55}}{14}=\frac{1\pm i\sqrt{55}}{14}$$

$$\frac{1}{14}-\frac{\sqrt{55}}{14}i \ \ \text{and} \ \ \frac{1}{14}+\frac{\sqrt{55}}{14}i$$

30. $\dfrac{-b\pm\sqrt{b^2-4ac}}{2a}$

$$\frac{-(-1)\pm\sqrt{(-1)^2-4(6)(5)}}{2(6)}=$$

$$\frac{1\pm\sqrt{-119}}{12}=\frac{1\pm i\sqrt{119}}{12}=$$

$$\frac{1}{12}-\frac{\sqrt{119}}{12}i \ \ \text{and} \ \ \frac{1}{12}+\frac{\sqrt{119}}{12}i$$

Chapter 10   Inequality

▶ When solving for X with an inequality sign, we assume the inequality sign is an equal sign and solve. Afterward, we replace the inequality sign. The inequality sign changes when we multiply or divide by a negative number on both sides; we flip the inequality sign.

Example 1:
Solve the inequality.

$$-5(2X - 5) + 7 > -18$$

$$-10X + 25 + 7 > -18$$

We assume the inequality sign is an equal sign and solve. We distribute first.

$$-10X + 32 > -18$$

We combine the like terms.

$$-10X > -50$$

We subtract 32 on both sides.

$$X < 5$$

We divide both sides by -10. We flip the inequality sign when we divide or multiply a negative number on both sides.

Our answer is $X < 5$.

» To solve for the inequality, we find the interval of x-coordinates that will make the inequality a true statement. For non-linear inequalities, we will create tables to solve the inequalities.

Example 2:
Solve the inequality.

$$X^2 + 8X + 15 < 0$$

We will create the table to solve the same inequality equation.

$$(X + 3)(X + 5) < 0$$
$$X = -3,\ X = -5$$

First, we find the zeros.

We create the intervals based on the zeros.

$$(-\infty, -5) \text{ and } (-5, -3) \text{ and } (-3, \infty)$$

$(-\infty, -5)$			
$(-5, -3)$			
$(-3, \infty)$			

We list the intervals in the first column.

	(X+3)	(X+5)	(X+3)(X+5)
$(-\infty, -5)$			
$(-5, -3)$			
$(-3, \infty)$			

We list the factors (X+3) and (X+5) and the inequality equation (X+3)(X+5) in the first row as shown left.

	(X+3)	(X+5)	(X+3)(X+5)
$(-\infty, -5)$	—	—	+
$(-5, -3)$	—	+	—
$(-3, \infty)$	+	+	+

We determine the signs (negative or positive) in each box. We pick any number in the interval and substitute the number in the factors to determine the signs. Then, we multiply the factors to identify the sign of the inequality equation in the given interval.

Solution: $(-5, -3)$

We had "less than" inequality, so we pick the interval that has a negative sign in the inequality equation column.

Example 3:
Solve the inequality.

$$(X - 2)^2 (X - 8) \ge 0$$

$$(X - 2)(X - 2)(X - 8) \ge 0$$

We find the zeros first.

$$X = 2, \; X = 8$$

We identify the intervals.

$$(-\infty, 2) \text{ and } (2, 8) \text{ and } (8, \infty)$$

	$(X-2)$	$(X-8)$	$(X-2)^2(X-8)$
$(-\infty, 2)$			
$(2, 8)$			
$(8, \infty)$			

We make the table.

	$(X-2)$	$(X-8)$	$(X-2)^2(X-8)$
$(-\infty, 2)$	$-$	$-$	$-$
$(2, 8)$	$+$	$-$	$-$
$(8, \infty)$	$+$	$+$	$+$

We determine the signs (negative or positive) in each box. We pick any number in the interval and substitute the number in the factors to identify the signs. Then, we substitute the signs in the inequality equation to determine the sign of the inequality equation in the given interval.

Solution: $[8, \infty) \cup \{2\}$

We had "greater than or equal to" inequality. We pick the interval that has positive sign in the inequality equation column. Because we have "equal to" part, we put a bracket around 8. We also have to include the number 2 as part of the solution because we set $X - 2 = 0$.

Example 4:

Solve the inequality

$$(X+5)(X+1)(X-8) > 0$$

$X = -5$, $X = -1$, $X = 8$

We find the zeros first.

$(-\infty, -5)$, $(-5, -1)$, $(-1, 8)$, and $(8, \infty)$

We identify the intervals.

116

	(X+5)	(X+1)	(X−8)	(X+5)(X+1)(X−8)
$(-\infty, -5)$	−	−	−	−
$(-5, -1)$	+	−	−	+
$(-1, 8)$	+	+	−	−
$(8, \infty)$	+	+	+	+

Solutions: $(-5, -1) \cup (8, \infty)$

Because of "greater than" inequality sign, we pick the interval with positive sign in the inequality equation column. We have two intervals and $\cup$ represents union, which indicates that the both intervals are the solutions.

Example 5:
Solve the inequality.

$$\frac{X-8}{X+2} \geq 0$$

We have a fraction. We set the numerator and denominator equal to zero, respectively. We solve for X.

$X - 8 = 0$ and $X + 2 = 0$
$X = 8$ and $X = -2$

$(-\infty, -2)$ and $(-2, 8)$ and $(8, \infty)$ We determine the intervals.

	X−8	X+2	$\frac{X-8}{X+2}$
$(-\infty, -2)$			
$(-2, 8)$			
$(8, \infty)$			

We make the table.

	X−8	X+2	$\dfrac{X-8}{X+2}$
$(-\infty, -2)$	−	−	+
$(-2, 8)$	−	+	−
$(8, \infty)$	+	+	+

We determine the signs (negative or positive) in each box. We pick any number in the interval and substitute the number in the factors to identify the signs. Then, we substitute the signs in the inequality equation to determine the sign of the inequality equation in the given interval.

Solution: $\left(-\infty, -2\right) \cup \left[8, \infty\right)$

We have a parenthesis, rather than a bracket, because it is undefined at X = −2 .

$$\frac{-2-8}{-2+2} = \frac{-10}{0} = \text{undefined}$$

Chapter 10   Inequality

1. Solve the inequality. $2(X+5)-7<27$	2. Solve the inequality. $5X-2\geq 23$
3. Solve the inequality. $7-12X+2(X-1)\leq 105$	4. Solve the inequality. $X^2+6X-55<0$
5. Solve the inequality. $-X^2-5X+14\geq 0$	6. Solve the inequality. $X^2-9X-10>0$
7. Solve the inequality. $-X^2-2X+35\leq 0$	8. Solve the inequality. $(X-2)^2(X+10)>0$
9. Solve the inequality. $(X-5)^2(X+14)<0$	10. Solve the inequality. $(2-X)(X-1)^2<0$

Chapter 10   Inequality

11. Solve the inequality. $(X+4)(X+2)^2 > 0$	12. Solve the inequality. $(X+11)^2(X-12) < 0$
13. Solve the inequality. $(X+1)(X-2)(X+4) > 0$	14. Solve the inequality. $(X-2)(X+5)(X-10) \le 0$
15. Solve the inequality. $(X-5)(3-X)(X+2) < 0$	16. Solve the inequality. $(X+2)(X-1)(X+5) < 0$
17. Solve the inequality. $(5-X)(X+5)(X+2) > 0$	18. Solve the inequality. $\dfrac{X+2}{X-4} > 0$
19. Solve the inequality. $\dfrac{X-1}{3-X} \le 0$	20. Solve the inequality. $\dfrac{X+4}{X-3} < 0$

Chapter 10   Inequality

21. Solve the inequality. $\dfrac{8-X}{X+2} > 0$	22. Solve the inequality. $\dfrac{X+1}{X-10} < 0$
23. Solve the inequality. $\dfrac{X-3}{X+7} \geq 0$	24. Solve the inequality. $(7-X)(X+4)(X-2) \geq 0$
25. Solve the inequality. $X^2 + 8X - 65 < 0$	26. Solve the inequality. $(X-4)^2(X+5) > 0$
27. Solve the inequality. $-X^2 + 16X + 17 > 0$	28. Solve the inequality. $(X-8)(X+1)(X+15) \leq 0$
29. Solve the inequality. $\dfrac{5-X}{X+15} < 0$	30. Solve the inequality. $(X+2)^2(X-1) > 0$

Chapter 10 Inequality
Answer key

1. $2(X+5)-7 < 27$

   $2X+10-7 < 27$

   $2X+3 < 27$

   $2X < 24$

   $X < 12$

2. $5X-2 \geq 23$

   $5X \geq 25$

   $X \geq 5$

3. $7-12X+2(X-1) \leq 105$

   $7-12X+2X-2 \leq 105$

   $-10X+5 \leq 105$

   $-10X \leq 100$

   $X \geq -10$

4. $X^2+6X-55 < 0$

   $(X+11)(X-5) < 0$

	$(X+11)$	$(X-5)$	$(X+1)(X-5)$
$(-\infty,-11)$	$-$	$-$	$+$
$(-11,5)$	$+$	$-$	$-$
$(5,\infty)$	$+$	$+$	$+$

Answer : $(-11,5)$

5. $-X^2-5X+14 \geq 0$

   $(X+7)(2-X) \geq 0$

	$(X+7)$	$(2-X)$	$(X+7)(2-X)$
$(-\infty,-7)$	$-$	$+$	$-$
$(-7,2)$	$+$	$+$	$+$
$(2,\infty)$	$+$	$-$	$-$

Answer: $[-7,2]$

6. $X^2 - 9X - 10 > 0$

   $(X+1)(X-10) > 0$

	$(X+1)$	$(X-10)$	$(X+1)(X-10)$
$(-\infty,-1)$	$-$	$-$	$+$
$(-1,10)$	$+$	$-$	$-$
$(10,\infty)$	$+$	$+$	$+$

Answer: $(-\infty,-1) \cup (10,\infty)$

7. $-X^2 - 2X + 35 \leq 0$

   $(X+7)(5-X) \leq 0$

	$(X+7)$	$(5-X)$	$(X+7)(5-X)$
$(-\infty,-7)$	$-$	$+$	$-$
$(-7,5)$	$+$	$+$	$+$
$(5,\infty)$	$+$	$-$	$-$

Answer: $(-\infty,-7] \cup [5,\infty)$

8. $(X-2)^2(X+10) > 0$

	$(X-2)$	$(X+10)$	$(X-2)^2(X+10)$
$(-\infty,-10)$	$-$	$-$	$-$
$(-10,2)$	$-$	$+$	$+$
$(2,\infty)$	$+$	$+$	$+$

Answer: $(-10,2) \cup (2,\infty)$

9. $(X-5)^2(X+14) < 0$

	$(X-5)$	$(X+14)$	$(X-5)^2(X+14)$
$(-\infty,-14)$	$-$	$-$	$-$
$(-14,5)$	$-$	$+$	$+$
$(5,\infty)$	$+$	$+$	$+$

Answer: $(-\infty,-14)$

10. $(2-X)(X-1)^2 < 0$

	$(2-X)$	$(X-1)$	$(2-X)(X-1)^2$
$(-\infty,1)$	$+$	$-$	$+$
$(1,2)$	$+$	$+$	$+$
$(2,\infty)$	$-$	$+$	$-$

Answer: $(2,\infty)$

11. $(X+4)(X+2)^2 > 0$

	$(X+4)$	$(X+2)$	$(X+4)(X+2)^2$
$(-\infty,-4)$	$-$	$-$	$-$
$(-4,-2)$	$+$	$-$	$+$
$(-2,\infty)$	$+$	$+$	$+$

Answer: $(-4,-2) \cup (-2,\infty)$

12. $(X+11)^2(X-12) < 0$

	$(X+11)$	$(X-12)$	$(X+11)^2(X-12)$
$(-\infty,-11)$	$-$	$-$	$-$
$(-11,12)$	$+$	$-$	$-$
$(12,\infty)$	$+$	$+$	$+$

Answer: $(-\infty,-11) \cup (-11,12)$

13. $(X+1)(X-2)(X+4) > 0$

	$(X+1)$	$(X-2)$	$(X+4)$	$(X+1)(X-2)(X+4)$
$(-\infty,-4)$	$-$	$-$	$-$	$-$
$(-4,-1)$	$-$	$-$	$+$	$+$
$(-1,2)$	$+$	$-$	$+$	$-$
$(2,\infty)$	$+$	$+$	$+$	$+$

Answer: $(-4,-1) \cup (2,\infty)$

14. $(X-2)(X+5)(X-10) \le 0$

	$(X-2)$	$(X+5)$	$(X-10)$	$(X-2)(X+5)(X-10)$
$(-\infty,-5)$	$-$	$-$	$-$	$-$
$(-5,2)$	$-$	$+$	$-$	$+$
$(2,10)$	$+$	$+$	$-$	$-$
$(10,\infty)$	$+$	$+$	$+$	$+$

Answer: $(-\infty,-5] \cup [2,10]$

15. $(X-5)(3-X)(X+2) < 0$

	$(X-5)$	$(3-X)$	$(X+2)$	$(X-5)(3-X)(X+2)$
$(-\infty,-2)$	$-$	$+$	$-$	$+$
$(-2,3)$	$-$	$+$	$+$	$-$
$(3,5)$	$-$	$-$	$+$	$+$
$(5,\infty)$	$+$	$-$	$+$	$-$

Answer: $(-2,3) \cup (5,\infty)$

16. $(X+2)(X-1)(X+5) < 0$

	$(X+2)$	$(X-1)$	$(X+5)$	$(X+2)(X-1)(X+5)$
$(-\infty,-5)$	$-$	$-$	$-$	$-$
$(-5,-2)$	$-$	$-$	$+$	$+$
$(-2,1)$	$+$	$-$	$+$	$-$
$(1,\infty)$	$+$	$+$	$+$	$+$

Answer: $(-\infty,-5) \cup (-2,1)$

17. $(5-X)(X+5)(X+2) > 0$

	$(5-X)$	$(X+5)$	$(X+2)$	$(5-X)(X+5)(X+2)$
$(-\infty,-5)$	$+$	$-$	$-$	$+$
$(-5,-2)$	$+$	$+$	$-$	$-$
$(-2,5)$	$+$	$+$	$+$	$+$
$(5,\infty)$	$-$	$+$	$+$	$-$

Answer: $(-\infty,-5) \cup (-2,5)$

18. $\dfrac{X+2}{X-4} > 0$

	$(X+2)$	$(X-4)$	$\dfrac{X+2}{X-4}$
$(-\infty,-2)$	$-$	$-$	$+$
$(-2,4)$	$+$	$-$	$-$
$(4,\infty)$	$+$	$+$	$+$

Answer: $(-\infty,-2) \cup (4,\infty)$

19. $\dfrac{X-1}{3-X} \leq 0$

	$(X-1)$	$(3-X)$	$\dfrac{X-1}{3-X}$
$(-\infty,1)$	$-$	$+$	$-$
$(1,3)$	$+$	$+$	$+$
$(3,\infty)$	$+$	$-$	$-$

Answer: $(-\infty,1] \cup (3,\infty)$

20. $\dfrac{X+4}{X-3} < 0$

	$(X+4)$	$(X-3)$	$\dfrac{X+4}{X-3}$
$(-\infty,-4)$	$-$	$-$	$+$
$(-4,3)$	$+$	$-$	$-$
$(3,\infty)$	$+$	$+$	$+$

Answer: $(-4,3)$

21. $\dfrac{8-X}{X+2} > 0$

	$(8-X)$	$(X+2)$	$\dfrac{8-X}{X+2}$
$(-\infty,-2)$	$+$	$-$	$-$
$(-2,8)$	$+$	$+$	$+$
$(8,\infty)$	$-$	$+$	$-$

Answer: $(-2,8)$

22. $\dfrac{X+1}{X-10} < 0$

	$(X+1)$	$(X-10)$	$\dfrac{X+1}{X-10}$
$(-\infty,-1)$	$-$	$-$	$+$
$(-1,10)$	$+$	$-$	$-$
$(10,\infty)$	$+$	$+$	$+$

Answer: $(-1,10)$

23. $\dfrac{X-3}{X+7} \geq 0$

	$(X-3)$	$(X+7)$	$\dfrac{X-3}{X+7}$
$(-\infty,-7)$	$-$	$-$	$+$
$(-7,3)$	$-$	$+$	$-$
$(3,\infty)$	$+$	$+$	$+$

Answer: $(-\infty,-7)\cup[3,\infty)$

24. $(7-X)(X+4)(X-2) \geq 0$

	$(7-X)$	$(X+4)$	$(X-2)$	$(7-X)(X+4)(X-2)$
$(-\infty,-4)$	+	−	−	+
$(-4,2)$	+	+	−	−
$(2,7)$	+	+	+	+
$(7,\infty)$	−	+	+	−

Answer: $(-\infty,-4] \cup [2,7]$

25. $X^2 + 8X - 65 < 0$
    $(X-5)(X+13) < 0$

	$(X-5)$	$(X+13)$	$(X-5)(X+13)$
$(-\infty,-13)$	−	−	+
$(-13,5)$	−	+	−
$(5,\infty)$	+	+	+

Answer: $(-13,5)$

26. $(X-4)^2(X+5) > 0$

	$(X-4)$	$(X+5)$	$(X-4)^2(X+5)$
$(-\infty,-5)$	−	−	−
$(-5,4)$	−	+	+
$(4,\infty)$	+	+	+

Answer: $(-5,4) \cup (4,\infty)$

27. $-X^2 + 16X + 17 > 0$
    $(17-X)(X+1) > 0$

	$(17-X)$	$(X+1)$	$(17-X)(X+1)$
$(-\infty,-1)$	+	−	−
$(-1,17)$	+	+	+
$(17,\infty)$	−	+	−

Answer: $(-1,17)$

28. $(X-8)(X+1)(X+15) \le 0$

	$(X-8)$	$(X+1)$	$(X+15)$	$(X-8)(X+1)(X+15)$
$(-\infty,-15)$	$-$	$-$	$-$	$-$
$(-15,-1)$	$-$	$-$	$+$	$+$
$(-1,8)$	$-$	$+$	$+$	$-$
$(8,\infty)$	$+$	$+$	$+$	$+$

Answer: $(-\infty,-15] \cup [-1,8]$

29. $\dfrac{5-X}{X+15} < 0$

	$(5-X)$	$(X+15)$	$\dfrac{5-X}{X+15}$
$(-\infty,-15)$	$+$	$-$	$-$
$(-15,5)$	$+$	$+$	$+$
$(5,\infty)$	$-$	$+$	$-$

Answer: $(-\infty,-15) \cup (5,\infty)$

30. $(X+2)^2(X-1) > 0$

	$(X+2)$	$(X-1)$	$(X+2)^2(X-1)$
$(-\infty,-2)$	$-$	$-$	$-$
$(-2,1)$	$+$	$-$	$-$
$(1,\infty)$	$+$	$+$	$+$

Answer: $(1,\infty)$

Chapter 11   Trigonometry Review, part 1

▶ We will first study converting from radian to degree and vice versa. Based on the measure $180° = \pi$, we create the proportion:

$$\frac{\pi}{180} = \frac{radian}{degree}$$

If we are converting from the radian to degree, we substitute the radian value in the numerator and solve for the denominator, the degree. From degree to radian, we substitute the degree value in the denominator and solve for the numerator, the radian.

Example 1:
Convert from 125° into radian.

$$\frac{\pi}{180} = \frac{radian}{degree}$$

$$\frac{\pi}{180} = \frac{X}{125} \qquad \text{We solve for X, which represents the radian.}$$

$$125\pi = 180X \qquad \text{We cross multiply and set the products equal.}$$

$$X = \frac{125\pi}{180} = \frac{25\pi}{36}$$

$$\frac{25\pi}{36} \text{ radian}$$

» We can also convert from degree to minutes or seconds. We have the measurement conversion: $1° = 60$ minutes and $1° = 3,600$ seconds. We symbolize minutes with ' and seconds with ". We set up proportions to convert between degree and minute or second.

$$\frac{1°}{60'} = \frac{degree}{minutes} \qquad \text{and} \qquad \frac{1°}{3,600"} = \frac{degree}{seconds}$$

Example 2:
Convert from 38.8° into minutes.

$$\frac{1°}{60'} = \frac{degree}{minutes}$$

$$\frac{1}{60} = \frac{38.8}{X} \qquad \text{We substitute the corresponding values and solve.}$$

$$X = (38.8)(60)$$

$$X = 2,328 \text{ minutes (or } 2,328')$$

» We can add or subtract degrees and minutes. The addition and subtraction rules apply the same. We can carry over or borrow a number. For example, when minutes go above 60, we can carry over 1°. Likewise, we can borrow 1°, which is equivalent to 60 minutes.

Example 3:
Simplify.

$$125° \ 15' - 37° \ 18' =$$

$15' - 18' =$ We subtract the minutes.

$(60' + 15') - 18' =$ We borrow 1°, which equals $60'$.
$75' - 18' = 57'$

$125° \rightarrow 124°$ The degree value decreases by 1 because of the borrowing.

$124° - 37° = 87°$ We subtract the degrees.

$$125° \ 15' - 37° \ 18' = 87° \ 57'$$

» Reference angle is the value of angle closest to the x-axis from the terminal side.

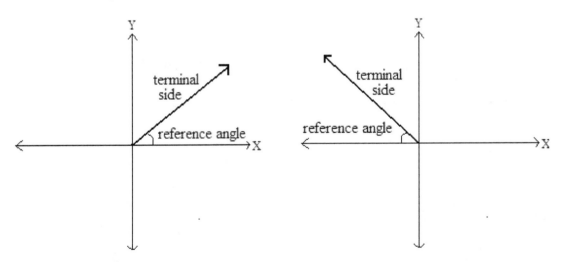

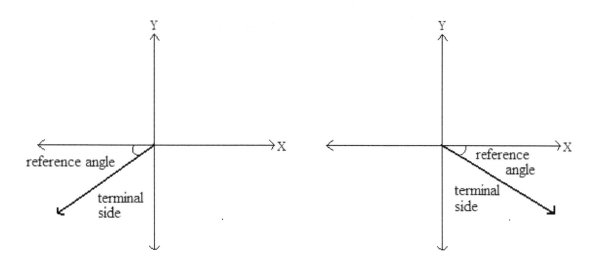

Example 4:
Find the reference angle of 215°.

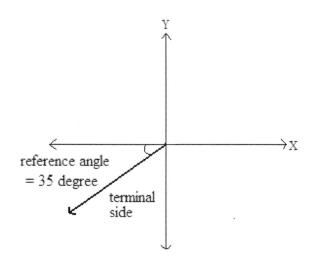

The terminal side is in the third quadrant because 215 degree is more than 180 degree but less than 270 degree. Now, we find the reference angle, which is the angle from the terminal side to the x-axis.

Once we determine the terminal side, we find the reference angle ($215-180$) in in this case. Therefore, our reference angle is 35°.

» We want to know the circular chart drawn below because the circular chart is helpful in finding the values of trigonometric values. For instance, the x-coordinate of the corresponding angle or radian is equal to cosine of that corresponding angle or radian. The y-coordinate is the sine. The tangent is y-coordinate divided by the x-coordinate.

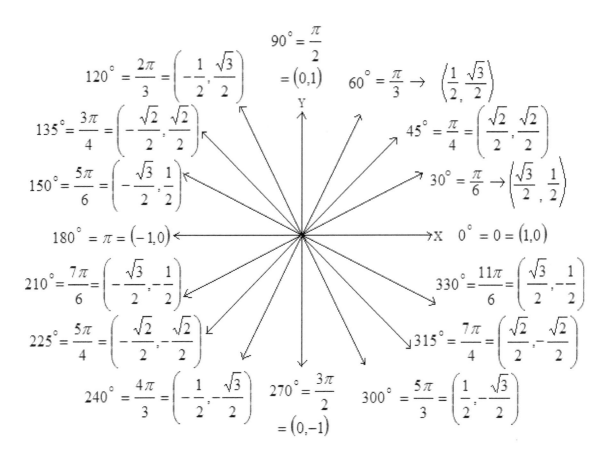

Example 5:
Without using a calculator, find the value.

$$\sin\left(\frac{3\pi}{4}\right) =$$

$$135° = \frac{3\pi}{4} = \left(-\frac{\sqrt{2}}{2}, \frac{\sqrt{2}}{2}\right)$$ According to the circular chart, we have this value.

$$\sin\left(\frac{3\pi}{4}\right) = \frac{\sqrt{2}}{2}$$ Sine is the y-coordinate value.

» If we have a right triangle, we can apply the Pythagorean theorem to find the unknown third side. Remember that the Pythagorean theorem cannot be applied to acute or obtuse triangle. In a right triangle, we have to have enough information to find the third side: either two lengths of the right triangle are given or two sides of the right triangle are equal. The Pythagorean theorem is $a^2 + b^2 = c^2$, where c is the hypotenuse.

Example 6:

Find the value of X.

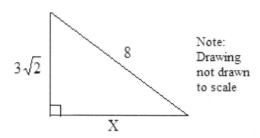

Note:
Drawing
not drawn
to scale

$a^2 + b^2 = c^2$ We apply the Pythagorean theorem.

$\left(3\sqrt{2}\right)^2 + X^2 = 8^2$ The hypotenuse (the side opposite the angle 90 degree) is 8.

$18 + X^2 = 64$ We solve for X.

$X^2 = 46$

$X = \sqrt{46}$

» Sine, cosine, and tangent have certain trigonometric ratios. To identify the ratio, we apply the "SOH-CAH-TOA" mnemonic.

Sine
Opposite
Hypotenuse
     $\text{Sine} = \dfrac{\text{Opposite}}{\text{Hypotenuse}}$

Cosine
Adjacent
Hypotenuse
     $\text{Cosine} = \dfrac{\text{Adjacent}}{\text{Hypotenuse}}$

Tangent
Opposite
Adjacent
     $\text{Tangent} = \dfrac{\text{Opposite}}{\text{Adjacent}}$

We also have Cosecant (csc), Secant (sec), and Cotangent (cot).

$$\text{Cosecant} = \text{csc} = \frac{1}{\sin} = \frac{1}{\left(\dfrac{opposite}{hypotenuse}\right)} = \frac{hypotenuse}{opposite}$$

$$\text{Secant} = \sec = \frac{1}{\cos} = \frac{1}{\left(\dfrac{adjacent}{hypotenuse}\right)} = \frac{hypotenuse}{adjacent}$$

$$\text{Cotangent} = \cot = \frac{1}{\tan} = \frac{1}{\left(\dfrac{opposite}{adjacent}\right)} = \frac{adjacent}{opposite}$$

Example 7:
Express the ratio of $\sec\theta$.

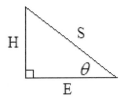

$$\sec\theta = \frac{hypotenuse}{adjacent}$$

$$\sec\theta = \frac{S}{E}$$

» There are two special triangles: 30°- 60°- 90° triangle and 45°- 45°- 90° triangle. We study these special triangles because the sides have special relations. If we know one of the lengths, then we can find the other two side lengths.

30°- 60°- 90° triangle

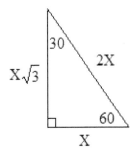

If opposite side of 30 degree angle is X, the hypotenuse is twice the value of X. The side opposite of 60 degree angle is $X\sqrt{3}$.

45°- 45°- 90° triangle

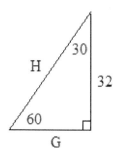

The sides opposite 45 degree angles are equal in length. If we represent the side as X, the hypotenuse is $X\sqrt{2}$.

Example 8:
Find the lengths of sides G and H.

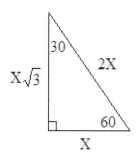

We think about the side length relations.

$$X\sqrt{3} = 32$$

The side length 32 corresponds with $X\sqrt{3}$. We set them equal and solve for X, which is the side length opposite the 30 degree angle.

$$X = \frac{32}{\sqrt{3}} = \frac{32\sqrt{3}}{3}$$

$$G = \frac{32\sqrt{3}}{3}$$

$$H = 2\left(\frac{32\sqrt{3}}{3}\right) = \frac{64\sqrt{3}}{3}$$

The length H is twice the length G.

» The law of sine has following relations. Based on the relations, we can find the unknown side or the unknown angle.

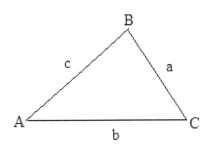

$$\frac{\sin A}{a} = \frac{\sin B}{b} = \frac{\sin C}{c} \qquad \text{or} \qquad \frac{a}{\sin A} = \frac{b}{\sin B} = \frac{c}{\sin C}$$

$$\frac{\sin (\text{measure of angle})}{\text{side opposite the angle}} = \frac{\sin (\text{measure of angle})}{\text{side opposite the angle}}$$

Example 9:

Find the length of side $\overline{EF}$.

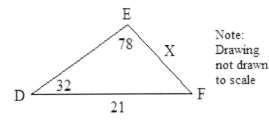

Note:
Drawing
not drawn
to scale

$$\frac{\sin A}{a} = \frac{\sin B}{b} = \frac{\sin C}{c}$$

$$\frac{\sin 78}{21} = \frac{\sin 32}{X} \qquad\qquad \text{We apply the law of sine.}$$

$$X \sin 78 = 21 \sin 32 \qquad\qquad \text{We solve for X.}$$

$$X = \frac{21 \sin 32}{\sin 78}$$

$$X \approx 11.38$$

» We also have the law of cosine.

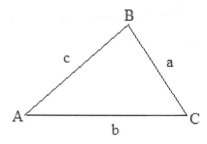

$$a^2 = b^2 + c^2 - 2bc(\cos A)$$
$$b^2 = a^2 + c^2 - 2ac(\cos B)$$
$$c^2 = a^2 + b^2 - 2ab(\cos C)$$

Applying the cosine formula helps us find the unknown side or unknown angle.

Example 10:
Find the measure of $\angle DEF$.

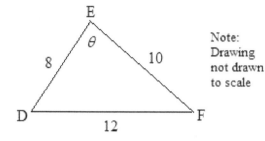

Note:
Drawing
not drawn
to scale

$$a^2 = b^2 + c^2 - 2bc(\cos A)$$

$$e^2 = d^2 + f^2 - 2df(\cos E)$$

$$12^2 = 10^2 + 8^2 - 2(10)(8)(\cos E)$$ We substitute the corresponding values.

$$144 = 100 + 64 - 160\cos E$$ We solve for the angle E.

$$-160\cos E = -20$$

$$\cos E = \frac{1}{8}$$

$$E = \cos^{-1}\left(\frac{1}{8}\right)$$

$$E \approx 82.82°$$

» If we have all the side lengths of a triangle, we can apply the Heron's formula to find the area of the triangle.

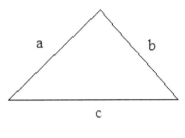

$$s = \frac{a+b+c}{2}$$

$$Area_{triangle} = \sqrt{s(s-a)(s-b)(s-c)}$$

The letter s represents the semi-perimeter of the triangle. We substitute the values of the semi-perimeter and side lengths into the formula to find the area of the triangle.

Example 11:
Applying Heron's formula, find the area of triangle.

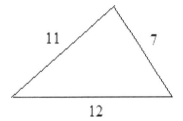

Note: Drawing not drawn to scale

$$s = \frac{a+b+c}{2} = \frac{11+7+12}{2} = 15 \qquad \text{We find the semi-perimeter first.}$$

$$s = 15,\ a = 11,\ b = 7,\ c = 12$$

$$
\begin{aligned}
Area_{triangle} &= \sqrt{s(s-a)(s-b)(s-c)} \\
&= \sqrt{(15)(15-11)(15-7)(15-12)} \\
&= \sqrt{1440} \\
&= 12\sqrt{10}
\end{aligned}
$$

We substitute the corresponding values into the equation.

Chapter 11   Trigonometry review, part I

1. Convert 84° into radian.	2. Convert $\dfrac{3\pi}{7}$ into degree.
3. Convert 124° into radian.	4. Convert $\dfrac{4\pi}{11}$ into degree.
5. Convert 425° into radian.	6. Convert $\dfrac{11\pi}{3}$ into degree.
7. Convert 14.5° into minutes.	8. Convert 84 minutes into degree.
9. Convert 32.7° into minutes.	10. Convert 148 minutes into degree.

Chapter 11   Trigonometry review, part I

11. Convert 24 degree into seconds.	12. Convert 10,000 seconds into degree.
13. Simplify. $58°\ 11' + 24°\ 24' =$	14. Simplify. $124°\ 28' - 58°\ 15' =$
15. Simplify. $125°\ 32' + 110°\ 21' =$	16. Simplify. $214°\ 31' - 151°\ 25' =$
17. Find the reference angle of $413°$ in degree.	18. Find the reference angle of $\dfrac{11\pi}{25}$ in radian.
19. Find the reference angle of $318°$ in degree.	20. Find the reference angle of $\dfrac{7\pi}{3}$ in radian.

Chapter 11   Trigonometry review, part I

21. Without using a calculator, find the value. $\sin\left(\dfrac{\pi}{3}\right) =$	22. Without using a calculator, find the value. $\cos\left(\dfrac{9\pi}{4}\right) =$
23. Without using a calculator, find the value. $\sin\left(\dfrac{3\pi}{2}\right) =$	24. Without using a calculator, find the value. $\cos\left(\dfrac{13\pi}{6}\right) =$
25. Find the value of *X*. Note: Drawing not drawn to scale 7 X $2\sqrt{3}$	26. Find the value of *X*. Note: Drawing not drawn to scale $5\sqrt{5}$ 21 X
27. Express the ratio of $\sin\theta$. 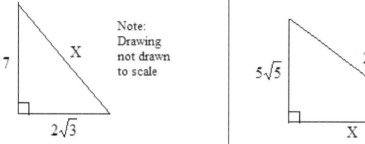	28. Express the ratio of $\tan\theta$. 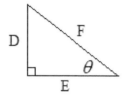
29. Express the ratio of $\sec\theta$. 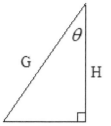	30. Express the ratio of $\cot\theta$. 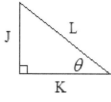

Chapter 11   Trigonometry review, part I

31. Find the lengths of sides G and H.	32. Find the lengths of sides G and H.
30 G H 60 5	11√2 45 H 45 G
33. Find the lengths of sides G and H.	**34.** Find the lengths of sides G and H.
30 H 14 60 G	45 G 25√5 45 H
35. Find the lengths of sides G and H.	**36.** Find the length of side $\overline{EF}$.
30 H 11√13 60 G	E 105 X 47 D 15 F Note: Drawing not drawn to scale
37. Find the length of side $\overline{DF}$.	**38.** Find the length of side $\overline{DF}$.
E 87 12 57 D X F Note: Drawing not drawn to scale	E 85 11 43 D X F Note: Drawing not drawn to scale
39. Find the length of side $\overline{EF}$.	**40.** Find the length of side $\overline{DF}$.
E 62 X 38 D 24 F Note: Drawing not drawn to scale	E 51 45 F 32 X D Note: Drawing not drawn to scale

Chapter 11   Trigonometry review, part I

41. Find the measure of ∠*DEF* . E θ 8 11 D 10 F Note: Drawing not drawn to scale	42. Find the length of side $\overline{DE}$. E X 5 D 7 42 F Note: Drawing not drawn to scale
43. Find the measure of ∠*DEF* . E θ 24 27 D 28 F Note: Drawing not drawn to scale	44. Find the length of side $\overline{EF}$. E 32 X 38 D 48 F Note: Drawing not drawn to scale
45. Find the measure of ∠*EDF* . E 48 52 θ D 35 F Note: Drawing not drawn to scale	46. Applying Heron's formula, find the area of triangle. 5 12 11 Note: Drawing not drawn to scale
47. Applying Heron's formula, find the area of triangle. 15 14 12 Note: Drawing not drawn to scale	48. Applying Heron's formula, find the area of triangle. 21 16 7 Note: Drawing not drawn to scale
49. Applying Heron's formula, find the area of triangle. 24 27 28 Note: Drawing not drawn to scale	50. Applying Heron's formula, find the area of triangle. 31 41 25 Note: Drawing not drawn to scale

Chapter 11 Trigonometry review, part I
Answer key

1. $\dfrac{\pi}{180} = \dfrac{X}{84}$
$180X = 84\pi$
$X = \dfrac{7\pi}{15}$

2. $\dfrac{\pi}{180} = \dfrac{\left(\dfrac{3\pi}{7}\right)}{X}$
$\pi X = \dfrac{540}{7}\pi$
$X = \dfrac{540}{7}$ or 77.14

3. $\dfrac{\pi}{180} = \dfrac{X}{124}$
$180X = 124\pi$
$X = \dfrac{31\pi}{45}$

4. $\dfrac{\pi}{180} = \dfrac{\left(\dfrac{4\pi}{11}\right)}{X}$
$\pi X = \dfrac{720}{11}\pi$
$X = \dfrac{720}{11}$ or 65.45

5. $\dfrac{\pi}{180} = \dfrac{X}{425}$
$180X = 425\pi$
$X = \dfrac{85\pi}{36}$

6. $\dfrac{\pi}{180} = \dfrac{\left(\dfrac{11\pi}{3}\right)}{X}$
$\pi X = 660\pi$
$X = 660\ degree$

7. $\dfrac{1}{60'} = \dfrac{14.5}{X}$
$X = 870$ minutes

8. $\dfrac{1}{60'} = \dfrac{X}{84'}$
$60X = 84$
$X = \dfrac{7}{5}$ or 1.4

9. $\dfrac{1}{60'} = \dfrac{32.7}{X}$
$X = 1962$ minutes

10. $\dfrac{1}{60'} = \dfrac{X}{148}$
$60X = 148$
$X = \dfrac{37}{15}$ or 2.47

11. $\dfrac{1}{3600''} = \dfrac{24}{X}$
$X = 86{,}400''$

12. $\dfrac{1}{3600''} = \dfrac{X}{10{,}000}$
$3600X = 10{,}000$
$X = \dfrac{25}{9}$ or 2.78

13. $82°\ 35'$

14. $66°\ 13'$

15. $235°\ 53'$

16. $63°\ 6'$

17. reference angle = 53 degree

18. reference angle = $\dfrac{11\pi}{25}$

19. reference angle = 42 degree

20. reference angle = $\dfrac{\pi}{3}$

21. $\sin\left(\dfrac{\pi}{3}\right) = \dfrac{\sqrt{3}}{2}$

22. $\cos\left(\dfrac{9\pi}{4}\right) = \dfrac{\sqrt{2}}{2}$

23. $\sin\left(\dfrac{3\pi}{2}\right) = -1$

24. $\cos\left(\dfrac{13\pi}{6}\right) = \dfrac{\sqrt{3}}{2}$

25. $(7)^2 + \left(2\sqrt{3}\right)^2 = X^2$

    $61 = X^2$

    $X = \sqrt{61}$

26. $\left(5\sqrt{5}\right)^2 + X^2 = 21^2$

    $125 + X^2 = 441$

    $X^2 = 316$

    $X = 2\sqrt{79}$

27. $\sin\theta = \dfrac{opposite}{hypotenuse}$

    $\sin\theta = \dfrac{B}{C}$

28. $\tan\theta = \dfrac{opposite}{adjacent}$

    $\tan\theta = \dfrac{D}{E}$

29. $\sec\theta = \dfrac{hypotenuse}{adjacent}$

    $\sec\theta = \dfrac{G}{H}$

30. $\cot\theta = \dfrac{adjacent}{opposite}$

    $\cot\theta = \dfrac{K}{J}$

31. $G = 5\sqrt{3}$
    $H = 10$

32. $G = 11$
    $H = 11$

33. $G = \dfrac{14\sqrt{3}}{3}$

    $H = \dfrac{28\sqrt{3}}{3}$

34. $G = \dfrac{25\sqrt{10}}{2}$

    $H = \dfrac{25\sqrt{10}}{2}$

35. $G = \dfrac{11\sqrt{39}}{3}$

    $H = \dfrac{22\sqrt{39}}{3}$

36. $\dfrac{\sin 105}{15} = \dfrac{\sin 47}{X}$

$X \sin 105 = 15 \sin 47$

$X = \dfrac{15 \sin 47}{\sin 105}$

$X = 11.36$

37. $\dfrac{\sin 57}{12} = \dfrac{\sin 87}{X}$

$X \sin 57 = 12 \sin 87$

$X = \dfrac{12 \sin 87}{\sin 57}$

$X = 14.29$

38. $\dfrac{\sin 43}{11} = \dfrac{\sin 85}{X}$

$X \sin 43 = 11 \sin 85$

$X = \dfrac{11 \sin 85}{\sin 43}$

$X = 16.07$

39. $\dfrac{\sin 62}{24} = \dfrac{\sin 38}{X}$

$X \sin 62 = 24 \sin 38$

$X = \dfrac{24 \sin 38}{\sin 62}$

$X = 16.73$

40. $\dfrac{\sin 45}{32} = \dfrac{\sin 51}{X}$

$X \sin 45 = 32 \sin 51$

$X = \dfrac{32 \sin 51}{\sin 45}$

$X = 35.17$

41. $10^2 = 8^2 + 11^2 - 2(8)(11)(\cos \theta)$

$100 = 64 + 121 - 176 \cos \theta$

$100 = 185 - 176 \cos \theta$

$-176 \cos \theta = -85$

$\cos \theta = \dfrac{85}{176}$

$\theta = \cos^{-1}\left(\dfrac{85}{176}\right) = 61.12°$

42. $X^2 = 5^2 + 7^2 - 2(5)(7)(\cos 42)$

$X^2 = 25 + 49 - 70 \cos 42$

$X^2 = 21.979$

$X = 4.6883$

43. $28^2 = 24^2 + 27^2 - 2(24)(27)(\cos \theta)$

$784 = 576 + 729 - 1296 \cos \theta$

$782 = 1305 - 1296 \cos \theta$

$-1296 \cos \theta = -521$

$\cos \theta = \dfrac{521}{1296}$

$\theta = \cos^{-1}\left(\dfrac{521}{1296}\right) = 66.3°$

44. $X^2 = 32^2 + 48^2 - 2(32)(48)(\cos 38)$

$X^2 = 1024 + 2304 - 3072 \cos 38$

$X^2 = 907.231$

$X = 30.12$

45. $52^2 = 48^2 + 35^2 - 2(48)(35)(\cos \theta)$

$2704 = 2304 + 1225 - 3360 \cos \theta$

$2704 = 3529 - 3360 \cos \theta$

$-3360 \cos \theta = -825$

$\cos \theta = \dfrac{55}{224}$

$\theta = \cos^{-1}\left(\dfrac{55}{224}\right) = 75.79°$

46. $s = \dfrac{a+b+c}{2} = \dfrac{5+12+11}{2} = 14$

$Area = \sqrt{s(s-a)(s-b)(s-c)}$

$\qquad = \sqrt{14(14-5)(14-12)(14-11)}$

$\qquad = 6\sqrt{21}$

(or approximately 27.5)

47. $s = \dfrac{a+b+c}{2} = \dfrac{15+14+12}{2} = 20.5$

$Area = \sqrt{s(s-a)(s-b)(s-c)}$

$= \sqrt{20.5(20.5-15)(20.5-14)(20.5-12)}$

$= \dfrac{\sqrt{99671}}{4}$

(or approximately 78.93)

48. $s = \dfrac{a+b+c}{2} = \dfrac{21+7+16}{2} = 22$

$Area = \sqrt{s(s-a)(s-b)(s-c)}$

$\qquad = \sqrt{22(22-21)(22-7)(22-16)}$

$\qquad = 6\sqrt{55}$

(or approximately 44.5)

49. $s = \dfrac{a+b+c}{2} = \dfrac{24+27+28}{2} = 39.5$

$Area = \sqrt{s(s-a)(s-b)(s-c)}$

$= \sqrt{39.5(39.5-24)(39.5-27)(39.5-28)}$

$= \dfrac{5\sqrt{56327}}{4}$

(or approximately 296.7)

50. $s = \dfrac{a+b+c}{2} = \dfrac{31+41+25}{2} = 48.5$

$Area = \sqrt{s(s-a)(s-b)(s-c)}$

$= \sqrt{48.5(48.5-31)(48.5-41)(48.5-25)}$

$= \dfrac{5\sqrt{95739}}{4}$

(or approximately 386.77)

Chapter 12   Trigonometry review, part II

▶ Arc length formula $s = r\theta$ helps us to find the partial distance around the circle. In the formula $s = r\theta$, s represents the arc length, r the radius, and θ the measure of angle in radian.

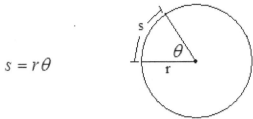

$$s = r\theta$$

Example 1:
Find the arc length.

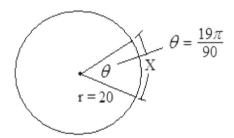

$\theta = \dfrac{19\pi}{90}$

$r = 20$

$s = r\theta$ We apply the arc length formula.

$$s = (20)\left(\frac{19\pi}{90}\right)$$

$$s = \frac{38\pi}{9}$$

» We can find the portion area of the circle by applying the area of sector formula $A = \dfrac{1}{2}r^2\theta$. The "r" represents the radius and the "θ" represents the measure of angle in radian.

The area of sector $= \dfrac{1}{2}r^2\theta$

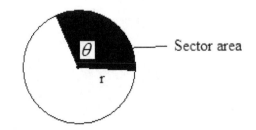

Sector area

Example 2:
Find the area of sector.

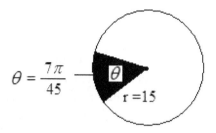

$$A = \frac{1}{2}r^2\theta$$

We apply the area of sector formula.

$$A = \frac{1}{2}(15)^2\left(\frac{7\pi}{45}\right)$$

We substitute the corresponding values.

$$A = \frac{35\pi}{2}$$

We have the area of sector.

» In the previous chapter, we studied the "SOH-CAH-TOA" mnemonic. We apply the mnemonic to find the unknown side. We find the side relations from the given angle measure. We identify sine, cosine, or tangent. Then, we substitute the corresponding values and evaluate the unknown side.

Example 3:
Find the value of X.

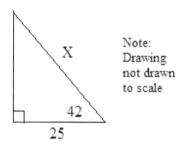

Note:
Drawing
not drawn
to scale

$$\cos = \frac{adjacent}{hypotenuse}$$

In respect to angle 42, we have adjacent and

hypotenuse sides. We have cosine.

$$\cos 42 = \frac{25}{X}$$

We substitute the corresponding values.

$$X\cos 42 = 25$$

We solve for X.

$$X = \frac{25}{\cos 42} \approx 33.64$$ We apply the calculator to find the answer.

» The amplitude shows the fluctuation of the graph. The higher the amplitude, the higher the graph will fluctuate. The amplitude is the half distance between the highest and lowest points on the graph. Hence, the higher the amplitude, the higher the fluctuation will be. The sine and cosine have amplitudes and the amplitude is the absolute value of the coefficient.

The absolute value of "a" is the amplitude.

$Y = a \sin bX$ $Y = a \cos bX$

   The amplitude $= |a|$

For cosecant and secant, we can find the range of y-values based on the value of "a". (As a side note, the cosecant is the reciprocal of the sine, and secant is the reciprocal of the cosine).

$Y = a \csc bX$ $Y = a \sec bX$

   Range: $Y \geq |a|$ and $Y \leq -|a|$

We do not have an amplitude for the tangent and cotangent (cotangent is the reciprocal of the tangent). The value "a" in tangent and cotangent indicates the rate of change. The higher the absolute value of "a", the faster the rate of change (the parts of the graph will become steeper).

$Y = a \tan bX$ $Y = a \cot bX$

   $|a| > 1$ The graph shows the faster rate of change.

   $0 < |a| < 1$ The graph shows the slower rate of change.

» The period indicates the interval that the graph repeats its major characteristics. We have the basic period 2π for sine, cosine, cosecant, and secant. We divide the period 2π by the value of "b" to obtain the period.

$Y = a \sin bX$ $Y = a \cos bX$ $Y = a \csc bX$ $Y = a \sec bX$

$$\text{Period} = \frac{2\pi}{b}$$

We have a different period for tangent and cotangent. The period of tangent or cotangent is π.

$$Y = a \tan bX \qquad\qquad Y = a \cot bX$$

$$\text{Period} = \frac{\pi}{b}$$

Example 4:

Find the amplitude and period

of $Y = -\dfrac{1}{5} \sin \dfrac{3}{7} X$.

$Y = a \sin bX$

Amplitude $= |a|$ We apply the amplitude formula.

$$\text{Amplitude} = \left| -\frac{1}{5} \right| = \frac{1}{5}$$

$$\text{Period} = \frac{2\pi}{b} \qquad\qquad \text{We apply the period formula.}$$

$$\text{Period} = \frac{2\pi}{\left(\dfrac{3}{7} \right)} = \frac{14\pi}{3}$$

Example 5:

Find the period and range

of $Y = 2 \sec \dfrac{1}{2} X$.

$Y = a \sec bX$

$$\text{Period} = \frac{2\pi}{b} \qquad\qquad \text{We apply the period formula.}$$

$$\text{Period} \frac{2\pi}{\left(\dfrac{1}{2} \right)} = 4\pi$$

Range: $Y \geq |a|$ and $Y \leq -|a|$ We apply the range formula.

$\quad\quad\quad Y \geq |2|$ and $Y \leq -|2|$

$\quad\quad\quad Y \geq 2$ and $Y \leq -2$

Example 6:

Find the period of $Y = 2\cot\dfrac{5\pi}{8}X$.

$$Y = a\cot bX$$

$$\text{Period} = \frac{\pi}{b} \quad\quad\quad \text{We apply the period formula.}$$

$$\text{Period} = \frac{\pi}{\left(\dfrac{5\pi}{8}\right)} = \frac{8}{5}$$

» The phase shift is the horizontal shift. If we have $f(X - h)$, the graph $f(X)$ shifts horizontally by h units. For $\sin(X - \pi)$, we have phase shift π (the graph shifts horizontally right by π units). We have to have X by itself inside the parenthesis to identify the phase shift. If we have $\cos(2X - \pi)$, we have to factor out 2 from the parenthesis to have X by itself. Hence, $\cos(2X - \pi)$ is same as $\cos 2\left(X - \dfrac{\pi}{2}\right)$; the phase shift is $\dfrac{\pi}{2}$.

Example 7:

Find the phase shift of $Y = \dfrac{1}{5}\sin\left(5X + \dfrac{\pi}{2}\right)$.

$$Y = \frac{1}{5}\sin 5\left(X + \frac{\pi}{10}\right) \quad\quad\quad \text{We factor out 5 to keep X by itself.}$$

$$\text{Phase shift: } -\frac{\pi}{10}$$

Another way to obtain the phase shift is to divide $-\dfrac{\pi}{2}$ by 5.

$$\text{Phase shift} = \frac{-\left(\dfrac{\pi}{2}\right)}{5} = -\frac{\pi}{10}$$

» We can find the measure of angle by taking inverse on both sides.

$\sin\theta = X$

$\sin^{-1}\sin\theta = \sin^{-1}X$ We take inverse sine on both sides to get the θ

$\theta = \sin^{-1}X$ by itself.

We can set up the ratio in the triangle to solve for the unknown angle. We apply the mnemonic SOH-CAH-TOA.

Example 8:
With calculator, find the angle θ.

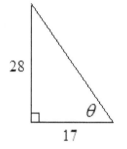

Note:
Drawing
not drawn
to scale

$\tan\theta = \dfrac{opposite}{adjacent} = \dfrac{28}{17}$ In respect to angle θ, tangent is equal to opposite side divided by adjacent side.

$\tan^{-1}\tan\theta = \tan^{-1}\dfrac{28}{17}$ To get θ by itself, we take inverse tangent on both sides of the equation.

$\theta = \tan^{-1}\dfrac{28}{17} \approx 58.74^{\circ}$ We apply the calculator to find the answer.

» We can apply the inverse method.

$\sin^{-1}\theta = X$

$\sin\sin^{-1}\theta = \sin X$ The inverse of $\sin^{-1}$ is $\sin$.

$\theta = \sin X$

Example 9:
Without calculator, find the value.

$$\cos\left(\sin^{-1}\left(\frac{\sqrt{5}}{5}\right)\right) =$$

$$\sin^{-1}\left(\frac{\sqrt{5}}{5}\right) = \theta \qquad\qquad \text{We set what goes in the parenthesis equal to } \theta.$$

$$\sin\sin^{-1}\left(\frac{\sqrt{5}}{5}\right) = \sin\theta \qquad\qquad \text{We take sine on both sides.}$$

$$\frac{\sqrt{5}}{5} = \sin\theta$$

$$\sin\theta = \frac{opposite}{hypotenuse} = \frac{\sqrt{5}}{5}$$

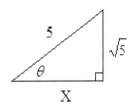

$$X^2 + \left(\sqrt{5}\right)^2 = 5^2 \qquad\qquad \text{We apply the Pythagorean theorem to find the X.}$$

$$X^2 + 5 = 25$$

$$X = 2\sqrt{5}$$

$$\cos\theta = \frac{adjacent}{hypotenuse} = \frac{2\sqrt{5}}{5}$$

$$\cos\left(\sin^{-1}\left(\frac{\sqrt{5}}{5}\right)\right) = \frac{2\sqrt{5}}{5}$$

» We have trigonometric identities. We can apply the identities to simplify equation.

$$\tan X = \frac{\sin X}{\cos X} \qquad \csc X = \frac{1}{\sin X} \qquad \sec X = \frac{1}{\cos X}$$

$$\cot X = \frac{1}{\tan X} = \frac{\cos X}{\sin X}$$

$$\sin^2 X + \cos^2 X = 1 \qquad \sec^2 X = \tan^2 X + 1 \qquad \csc^2 X = \cot^2 X + 1$$

Example 10:
Simplify.

$$\frac{8\sqrt{2} \sin^2 X + 8\sqrt{2} \cos^2 X}{5 \cot^2 X + 5} =$$

$$\frac{8\sqrt{2}\left(\sin^2 X + \cos^2 X\right)}{5\left(\cot^2 X + 1\right)} =$$ We factor the numerator and the denominator. We see that we have two identities.

$$\frac{8\sqrt{2}(1)}{5\left(\csc^2 X\right)} =$$ We replace the trigonometric identities.

$$\frac{8\sqrt{2}}{\left(\dfrac{5}{\sin^2 X}\right)} =$$ We flip the denominator.

$$\frac{8\sqrt{2} \sin^2 X}{5}$$

» We will study the sum and difference formulas. The formulas help us to calculate the values without a calculator. The formulas will not work in all cases, but the ones that have the angle values on the circular chart, we will have the answers.

-Sum formulas:

$$\sin(A + B) = \sin A \cos B + \cos A \sin B$$
   (e.g. $\sin(45 + 30) = \sin 45 \cos 30 + \cos 45 \sin 30$)

$$\cos(A + B) = \cos A \cos B - \sin A \sin B$$
   (e.g. $\cos(120 + 45) = \cos 120 \cos 45 - \sin 120 \sin 45$)

$$\tan(A+B) = \frac{\tan A + \tan B}{1 - \tan A \tan B}$$

$$\text{(e.g. } \tan(45 + 60) = \frac{\tan 45 + \tan 60}{1 - \tan 45 \tan 60}\text{)}$$

-Difference formulas:

$$\sin(A-B) = \sin A \cos B - \cos A \sin B$$

$$\text{(e.g. } \sin(60 - 45) = \sin 60 \cos 45 - \cos 60 \sin 45\text{)}$$

$$\cos(A-B) = \cos A \cos B + \sin A \sin B$$

$$\text{(e.g. } \cos(45 - 30) = \cos 45 \cos 30 + \sin 45 \sin 30\text{)}$$

$$\tan(A-B) = \frac{\tan A - \tan B}{1 + \tan A \tan B}$$

$$\text{(e.g. } \tan(120 - 45) = \frac{\tan 120 - \tan 45}{1 + \tan 120 \tan 45}\text{)}$$

Example 11:
Without a calculator, find the value.

$$\sin 28 \cos 32 + \cos 28 \sin 32 =$$

$$\sin(A+B) = \sin A \cos B + \cos A \sin B \qquad \text{We apply the sum formula.}$$

$$A = 28, \ B = 32 \qquad \text{We identify } A \text{ and } B \text{ values.}$$

$$\sin 28 \cos 32 + \cos 28 \sin 32 = \sin(28 + 32) = \sin 60$$

$$60° \rightarrow \left(\frac{1}{2}, \frac{\sqrt{3}}{2}\right); \ \sin 60 = \frac{\sqrt{3}}{2}$$

$$\sin 28 \cos 32 + \cos 28 \sin 32 = \frac{\sqrt{3}}{2}$$

Example 12:
Without a calculator, find the value.

$$\sin(15) =$$

$$\sin(A-B) = \sin A \cos B - \cos A \sin B \qquad \text{We apply the difference formula.}$$

$$\sin(45 - 30) = \sin 45 \cos 30 - \cos 45 \sin 30$$

$$45° \rightarrow \left(\frac{\sqrt{2}}{2}, \frac{\sqrt{2}}{2} \right) \quad \text{and} \quad 30° \rightarrow \left(\frac{\sqrt{3}}{2}, \frac{1}{2} \right)$$

$$= \left(\frac{\sqrt{2}}{2} \right)\left(\frac{\sqrt{3}}{2} \right) - \left(\frac{\sqrt{2}}{2} \right)\left(\frac{1}{2} \right) = \frac{\sqrt{6}}{4} - \frac{\sqrt{2}}{4}$$

$$\sin(15) = \frac{\sqrt{6}}{4} - \frac{\sqrt{2}}{4} = \frac{\sqrt{6} - \sqrt{2}}{4}$$

» We also have half angle formulas. We apply the formulas to find the values without applying the calculator.

$$\sin \frac{A}{2} = \pm \sqrt{\frac{1 + \cos A}{2}}$$

$$\cos \frac{A}{2} = \pm \sqrt{\frac{1 + \cos A}{2}}$$

$$\tan \frac{A}{2} = \frac{\sin A}{1 + \cos A}$$

Example 13:
Without calculator, find the value.

$$\tan 22.5 =$$ If we double the angle 22.5, then we obtain 45. Hence, we apply the half-angle formula.

$$\tan 22.5 = \tan \frac{45}{2}$$ The A is 45.

$$\tan \frac{A}{2} = \frac{\sin A}{1 + \cos A}$$ We apply the half-angle formula.

$$\tan \frac{45}{2} = \frac{\sin 45}{1 + \cos 45}$$

$$45° \rightarrow \left(\frac{\sqrt{2}}{2}, \frac{\sqrt{2}}{2} \right)$$

$$= \frac{\frac{\sqrt{2}}{2}}{1+\frac{\sqrt{2}}{2}} = \frac{\frac{\sqrt{2}}{2}}{\frac{2+\sqrt{2}}{2}} = \left(\frac{\sqrt{2}}{2}\right)\left(\frac{2}{2+\sqrt{2}}\right)$$

$$= \frac{\sqrt{2}}{2+\sqrt{2}} = \frac{\sqrt{2}}{2+\sqrt{2}}\left(\frac{2-\sqrt{2}}{2-\sqrt{2}}\right)$$ We rationalize by multiplying conjugates.

$$= \frac{2\sqrt{2}-2}{4-2} = \frac{2\sqrt{2}-2}{2} = \sqrt{2}-1$$

$$\tan 22.5 = \sqrt{2}-1$$

» We graphed applying two variables. In two dimension graphs, we can introduce a third variable by applying the parametric equations. Given the values of X and Y, we can represent the third variables, such as T, to represent the values of T at X and Y. Traditional example is to represent the T as time. At given points of X and Y, we can determine the time T. Given the value of T, we will determine the values of X and Y by completing the table.

Example 14:
Complete the table.

$$X = -2t - 5 \quad Y = 8t + 2$$

These are parametric equations. We will substitute the *t* values into each equation to find the values of X and Y.

t	X	Y
−5		
0		
2		

$$X = -2(-5) - 5 = 5$$
$$Y = 8(-5) + 2 = -38$$

For $t = -5$, we substitute the number into each equation to find the values of X and Y. Then, we write the values in the table.

t	X	Y
−5	5	−38
0	−5	2
2	−9	18

» We can combine the two parametric equations to create an equation in terms of X and Y. We set one parametric equation equal to *t* and substitute the *t* into the other parametric equation. We will have an equation in terms of X and Y, and we can graph the equation on a X-Y plane.

Example 15:
Express the equation in terms of X and Y.

$$Y = 2t + 5 \qquad\qquad X = \frac{t-2}{5}$$

$$X = \frac{t-2}{5} \quad \rightarrow \quad t = 5X + 2$$ We get the *t* by itself (In this case, we will work with the second equation).

$$t = 5X + 2$$
$$\downarrow$$
$$Y = 2t + 5$$ We substitute the second equation into the first equation.

$$Y = 2(5X + 2) + 5$$ Then, we simplify.
$$Y = 10X + 4 + 5$$
$$Y = 10X + 9$$

» Polar coordinate is represented by (r, θ). The r is the length and the θ is the measure of the angle. We can convert from the rectangular coordinate to polar coordinate and vice versa.

To convert from the rectangular coordinate (X, Y) to polar coordinate (r, θ), we apply the following formula.

$$r = \pm\sqrt{X^2 + Y^2} \qquad \text{and} \qquad \tan\theta = \frac{Y}{X}$$

Example 16:
Convert from rectangular coordinates
to polar coordinates.

$$(1, 5)$$ We have the rectangular coordinate. The value of X is 1

and the value of Y is 5.

$r = \pm\sqrt{X^2 + Y^2}$ We apply the formula.

$r = \pm\sqrt{(1)^2 + (5)^2}$
$= \sqrt{1+25} = \sqrt{26}$ The length is $\sqrt{26}$.

$\tan\theta = \dfrac{Y}{X}$

$\tan\theta = \dfrac{5}{1}$

$\tan^{-1}\tan\theta = \tan^{-1} 5$ We take inverse on both sides.
$\theta = \tan^{-1} 5$
$\theta = 78.7\,°$ The measure of degree is 78.7°.

$(\sqrt{26}, 78.7°)$ We have the polar coordinate.

» We can also convert the polar coordinate to the rectangular coordinate. Given the polar coordinate (r,θ), we can find the rectangular coordinate (X,Y).

$X = r\cos\theta$
$Y = r\sin\theta$

Example 17:
Convert from polar coordinate to
rectangular coordinate.

$(2\sqrt{3}, 25°)$

$X = r\cos\theta$ We apply the formula.
$= (2\sqrt{3})(\cos 25) = 3.14$

$Y = r\sin\theta$
$= (2\sqrt{3})(\sin 25) = 1.46$

$(3.14, 1.46)$ We have the rectangular coordinate.

Chapter 12   Trigonometry review, part II

1. Find the arc length. 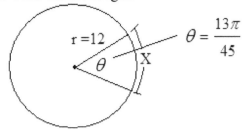 $r = 12$ $\theta = \dfrac{13\pi}{45}$	2. Find the arc length. 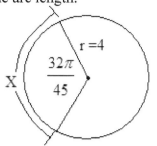 $r = 4$ $\dfrac{32\pi}{45}$
3. Find the arc length. 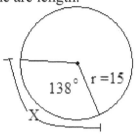 $138°$ $r = 15$	4. Find the area of sector. 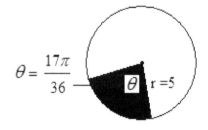 $\theta = \dfrac{17\pi}{36}$ $r = 5$
5. Find the area of sector. 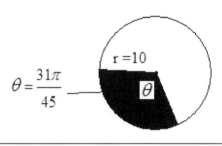 $r = 10$ $\theta = \dfrac{31\pi}{45}$	6. Find the area of sector. 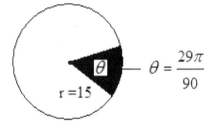 $r = 15$ $\theta = \dfrac{29\pi}{90}$
7. Find the value of X. 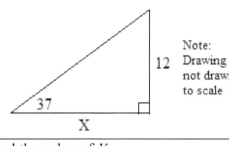 12 37 X Note: Drawing not drawn to scale	8. Find the value of X. 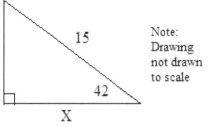 15 42 X Note: Drawing not drawn to scale
9. Find the value of X. 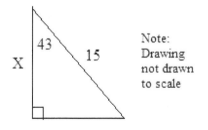 43 15 X Note: Drawing not drawn to scale	10. Find the value of X. 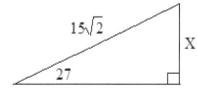 $15\sqrt{2}$ X 27 Note: Drawing not drawn to scale

Chapter 12   Trigonometry review, part II

11. Find the amplitude and period of $Y = 5\sin\frac{1}{8}X$	12. Find the amplitude and period of $Y = \frac{1}{7}\cos\frac{1}{2}X$
13. Find the amplitude and period of $Y = -2\sin 2X$	14. Find the period of $Y = \tan 3X$
15. Find the period of $Y = \frac{1}{2}\tan\frac{\pi}{7}X$	16. Find the phase shift of $Y = -3\sin\left(X + \frac{\pi}{4}\right)$
17. Find the phase shift of $Y = \frac{1}{5}\sin\left(3X - \frac{1}{2}\right)$	18. Find the phase shift of $Y = -2\cos\left(5X + \frac{4\pi}{7}\right)$
19. Find the period and range of $Y = 8\csc\frac{1}{3}X$	20. Find the period and range of $Y = \frac{-1}{4}\sec\frac{2\pi}{5}X$

Chapter 12   Trigonometry review, part II

21. Find the period and range of $Y = -3\csc\dfrac{\pi}{8}X$	22. Find the period and range of $Y = \dfrac{1}{5}\sec 5X$
23. Find the period of $Y = 3\cot\dfrac{2\pi}{7}X$	24. Find the period of $Y = \dfrac{1}{2}\cot 12X$
25. Find the period of $Y = \dfrac{-1}{4}\cot\dfrac{4\pi}{5}X$	26. With calculator, find the angle θ. 17 16 Note: Drawing not drawn to scale θ
27. With calculator, find the angle θ. 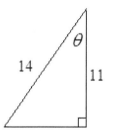 θ 14 11 Note: Drawing not drawn to scale	28. With calculator, find the angle θ. 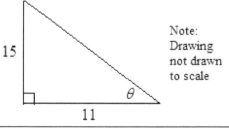 15 Note: Drawing not drawn to scale θ 11
29. Without calculator, find the value ($\sin\theta > 0$). $\sin\left(\cos^{-1}\dfrac{2}{5}\right) =$	30. Without calculator, find the value ($\tan\theta > 0$). $\tan\left(\sin^{-1}\dfrac{1}{5}\right) =$

Chapter 12   Trigonometry review, part II

31. Without calculator, find the value ($\cos\theta > 0$). $\cos\left(\tan^{-1}\dfrac{\sqrt{3}}{5}\right) =$	32. Simplify. $5\sqrt{2}\sin^2\theta + 5\sqrt{2}\cos^2\theta =$
33. Simplify. $\tan^2\theta\cos^2\theta + \cot^2\theta\sin^2\theta =$	34. Simplify. $\left(5\sin^2\theta + 5\cos^2\theta\right)\left(4\cos^2\theta\right) =$
35. Simplify. $\dfrac{7\sqrt{2}\sin^2\theta + 7\sqrt{2}\cos^2\theta}{7 + 7\tan^2\theta} =$	36. Simplify. $2\sin^2\theta + \tan^2\theta + 2\cos^2\theta + 1 =$
37. Without calculator, find the value. $\cos 37\cos 23 - \sin 37\sin 23 =$	38. Without calculator, find the value. $\sin\dfrac{17\pi}{24}\cos\dfrac{5\pi}{8} + \cos\dfrac{17\pi}{24}\sin\dfrac{5\pi}{8} =$
39. Without calculator, find the value. $\dfrac{\tan 78 - \tan 33}{1 + (\tan 78)(\tan 33)} =$	40. Without calculator, find the value. $\sin 75 =$

Chapter 12   Trigonometry review, part II

41. Without calculator, find the value. $\tan 112.5 =$	42. Without calculator, find the value. $\sin 67.5 =$

43. Without calculator, find the value.

$\cos 15 =$

44. Complete the table.
   $X = 2t + 7$ and $Y = 8t - 1$

t	X	Y
−2		
0		
5		

45. Complete the table.
   $X = -5t + 2$ and $Y = 3t - 3$

t	X	Y
−5		
−3		
4		

46. Express the equation in terms of X and Y.

$Y = 3t - 7$ and $X = 3t + 1$

47. Express the equation in terms of X and Y. $Y = 2t + 7$ and $X = t - 1$	48. Convert from polar coordinates to rectangular coordinates. $(7, 37°)$

49. Convert from rectangular coordinates to polar coordinates. $(5, -4)$	50. Convert from polar coordinates to rectangular coordinates. $(5\sqrt{2}, 48.3°)$

Chapter 12 Trigonometry review, part II
Answer key

1. $s = r\theta$

$$s = (12)\left(\frac{13\pi}{45}\right) = \frac{52\pi}{15}$$

2. $s = r\theta$

$$s = (4)\left(\frac{32\pi}{45}\right) = \frac{128\pi}{45}$$

3. $\dfrac{\pi}{180} = \dfrac{\theta}{138}$

$$\theta = \frac{138\pi}{180} = \frac{23\pi}{30}$$

$$s = r\theta$$

$$s = (15)\left(\frac{23\pi}{30}\right) = \frac{23\pi}{2}$$

4. $Area = \dfrac{1}{2}r^2\theta = \dfrac{1}{2}(5)^2\left(\dfrac{17\pi}{36}\right) = \dfrac{425\pi}{72}$

5.
$$Area = \frac{1}{2}r^2\theta = \frac{1}{2}(10)^2\left(\frac{31\pi}{45}\right) = \frac{310\pi}{9}$$

6.
$$Area = \frac{1}{2}r^2\theta = \frac{1}{2}(15)^2\left(\frac{29\pi}{90}\right) = \frac{145\pi}{4}$$

7. $\tan 37 = \dfrac{12}{X}$

$$X \tan 37 = 12$$

$$X = \frac{12}{\tan 37} = 15.92$$

8. $\cos 42 = \dfrac{X}{15}$

$$X = 15\cos 42 = 11.15$$

9. $\cos 43 = \dfrac{X}{15}$

$$X = 15\cos 43 = 10.97$$

10. $\sin 27 = \dfrac{X}{15\sqrt{2}}$

$$X = (15\sqrt{2})(\sin 27) = 9.63$$

11. $amplitude = 5$

$$period = \frac{2\pi}{\left(\frac{1}{8}\right)} = 16\pi$$

12. $amplitude = \dfrac{1}{7}$

$$period = \frac{2\pi}{\left(\frac{1}{2}\right)} = 4\pi$$

13. $amplitude = |-2| = 2$

$$period = \frac{2\pi}{2} = \pi$$

14. $period = \dfrac{\pi}{3}$

15. $period = \dfrac{\pi}{\left(\dfrac{\pi}{7}\right)} = 7$

16. $phase.shift = \dfrac{-\pi}{4}$

17. $phase.shift = \dfrac{\left(\dfrac{1}{2}\right)}{3} = \dfrac{1}{6}$

18. $phase.shift = \dfrac{\left(\dfrac{-4\pi}{7}\right)}{5} = \dfrac{-4\pi}{35}$

19. $period = \dfrac{2\pi}{\left(\dfrac{1}{3}\right)} = 6\pi$

    Range: $Y \geq |8|$ and $Y \leq -|8|$

    $\qquad Y \geq 8$ and $Y \leq -8$

20. $period = \dfrac{2\pi}{\left(\dfrac{2\pi}{5}\right)} = 5$

    Range: $Y \geq \left|\dfrac{-1}{4}\right|$ and $Y \leq -\left|\dfrac{-1}{4}\right|$

    $\qquad Y \geq \dfrac{1}{4}$ and $Y \leq -\dfrac{1}{4}$

21. $period = \dfrac{2\pi}{\left(\dfrac{\pi}{8}\right)} = 16$

    Range: $Y \geq |-3|$ and $Y \leq -|-3|$

    $\qquad Y \geq 3$ and $Y \leq -3$

22. $period = \dfrac{2\pi}{5}$

    range: $Y \geq \left|\dfrac{1}{5}\right|$ and $Y \leq -\left|\dfrac{1}{5}\right|$

    $\qquad Y \geq \dfrac{1}{5}$ and $Y \leq -\dfrac{1}{5}$

23. $period = \dfrac{\pi}{\left(\dfrac{2\pi}{7}\right)} = \dfrac{7}{2}$

24. $period = \dfrac{\pi}{12}$

25. $period = \dfrac{\pi}{\left(\dfrac{4\pi}{5}\right)} = \dfrac{5}{4}$

26. $\sin\theta = \dfrac{16}{17}$

    $\theta = \sin^{-1}\left(\dfrac{16}{17}\right) = 70.25°$

27. $\cos\theta = \dfrac{11}{14}$

    $\theta = \cos^{-1}\left(\dfrac{11}{14}\right) = 38.21°$

28. $\tan\theta = \dfrac{15}{11}$

    $\theta = \tan^{-1}\left(\dfrac{15}{11}\right) = 53.75°$

29. $\cos^{-1}\dfrac{2}{5} = \theta$

    $\cos\theta = \dfrac{2}{5}$

    $2^2 + X^2 = 5^2$

    $X = \sqrt{21}$

    $\sin\left(\cos^{-1}\dfrac{2}{5}\right) = \dfrac{\sqrt{21}}{5}$

30. $\sin^{-1}\dfrac{1}{5} = \theta$

$\sin\theta = \dfrac{1}{5}$

$X^2 + 1^2 = 5^2$

$X = 2\sqrt{6}$

$\tan\left(\sin^{-1}\dfrac{1}{5}\right) = \dfrac{1}{2\sqrt{6}} = \dfrac{\sqrt{6}}{12}$

31. $\tan^{-1}\left(\dfrac{\sqrt{3}}{5}\right) = \theta$

$\tan\theta = \dfrac{\sqrt{3}}{5}$

$(5)^2 + \left(\sqrt{3}\right)^2 = X^2$

$X = 2\sqrt{7}$

$\cos\left(\tan^{-1}\dfrac{\sqrt{3}}{5}\right) = \dfrac{5}{2\sqrt{7}} = \dfrac{5\sqrt{7}}{14}$

32. $5\sqrt{2}\sin^2\theta + 5\sqrt{2}\cos^2\theta =$
$5\sqrt{2}\left(\sin^2\theta + \cos^2\theta\right) =$
$\left(5\sqrt{2}\right)(1) = 5\sqrt{2}$

33. $\tan^2\theta\cos^2\theta + \cot^2\theta\sin^2\theta =$
$\left(\dfrac{\sin^2\theta}{\cos^2\theta}\right)\cos^2\theta + \dfrac{\cos^2\theta}{\sin^2\theta}\sin^2\theta =$
$\sin^2\theta + \cos^2\theta = 1$

34. $\left(5\sin^2\theta + 5\cos^2\theta\right)\left(4\cos^2\theta\right) =$
$\left(5\left(\sin^2\theta + \cos^2\theta\right)\right)\left(4\cos^2\theta\right) =$
$\left(5(1)\right)\left(4\cos^2\theta\right) = 20\cos^2\theta$

35. $\dfrac{7\sqrt{2}\sin^2\theta + 7\sqrt{2}\cos^2\theta}{7 + 7\tan^2\theta} =$

$\dfrac{7\sqrt{2}\left(\sin^2\theta + \cos^2\theta\right)}{7\left(1 + \tan^2\theta\right)} = \dfrac{\left(\sqrt{2}\right)(1)}{\sec^2\theta}$

$= \dfrac{\sqrt{2}}{\sec^2 X} = \dfrac{\sqrt{2}}{\left(\dfrac{1}{\cos^2 X}\right)} = \sqrt{2}\cos^2 X$

36. $2\sin^2\theta + \tan^2\theta + 2\cos^2\theta + 1 =$
$2\sin^2\theta + 2\cos^2\theta + \tan^2\theta + 1 =$
$2\left(\sin^2\theta + \cos^2\theta\right) + \sec^2\theta =$
$2(1) + \sec^2\theta$
$= 2 + \sec^2\theta$

37. $\cos 37\cos 23 - \sin 37\sin 23 =$
$\cos(37 + 23) = \cos 60$

$60° \rightarrow \left(\dfrac{1}{2}, \dfrac{\sqrt{3}}{2}\right)$ and $\cos 60 = \dfrac{1}{2}$

$\cos 37\cos 23 - \sin 37\sin 23 = \dfrac{1}{2}$

38. $\sin\dfrac{17\pi}{24}\cos\dfrac{5\pi}{8} + \cos\dfrac{17\pi}{24}\sin\dfrac{5\pi}{8} =$

$\sin\left(\dfrac{17\pi}{24} + \dfrac{5\pi}{8}\right) = \sin\left(\dfrac{4\pi}{3}\right)$

$\dfrac{4\pi}{3} \rightarrow \left(\dfrac{-1}{2}, \dfrac{-\sqrt{3}}{2}\right)$

$\sin\left(\dfrac{4\pi}{3}\right) = \dfrac{-\sqrt{3}}{2}$

$\sin\dfrac{17\pi}{24}\cos\dfrac{5\pi}{8} + \cos\dfrac{17\pi}{24}\sin\dfrac{5\pi}{8} = \dfrac{-\sqrt{3}}{2}$

39. $\dfrac{\tan 78 - \tan 33}{1 + (\tan 78)(\tan 33)} =$

$\tan(78 - 33) = \tan 45$

$45° \rightarrow \left(\dfrac{\sqrt{2}}{2}, \dfrac{\sqrt{2}}{2}\right)$

$\tan 45 = \dfrac{\sqrt{2}}{2} \div \dfrac{\sqrt{2}}{2} = 1$

$\dfrac{\tan 78 - \tan 33}{1 + (\tan 78)(\tan 33)} = 1$

40. $\sin 75 = \sin(45 + 30) =$

$\sin 45 \cos 30 + \cos 45 \sin 30$

$45° \rightarrow \left(\dfrac{\sqrt{2}}{2}, \dfrac{\sqrt{2}}{2}\right)$ and $30° \rightarrow \left(\dfrac{\sqrt{3}}{2}, \dfrac{1}{2}\right)$

$\left(\dfrac{\sqrt{2}}{2}\right)\left(\dfrac{\sqrt{3}}{2}\right) + \left(\dfrac{\sqrt{2}}{2}\right)\left(\dfrac{1}{2}\right) =$

$\dfrac{\sqrt{6}}{4} + \dfrac{\sqrt{2}}{4} = \dfrac{\sqrt{6} + \sqrt{2}}{4}$

41. $\tan 112.5 = \tan \dfrac{225}{2} = \dfrac{\sin 225}{1 + \cos 225} =$

$\dfrac{\dfrac{-\sqrt{2}}{2}}{1 + \left(\dfrac{-\sqrt{2}}{2}\right)} = \dfrac{\dfrac{-\sqrt{2}}{2}}{\left(\dfrac{2 - \sqrt{2}}{2}\right)} =$

$\dfrac{-\sqrt{2}}{2}\left(\dfrac{2}{2 - \sqrt{2}}\right) = \dfrac{-\sqrt{2}}{2 - \sqrt{2}} = -\sqrt{2} - 1$

42. $\sin 67.5 = \sin \dfrac{135}{2} =$

$\sqrt{\dfrac{1 - \cos 135}{2}} = \sqrt{\dfrac{1 - \left(\dfrac{-\sqrt{2}}{2}\right)}{2}} = \sqrt{\dfrac{1 + \dfrac{\sqrt{2}}{2}}{2}}$

$= \sqrt{\dfrac{\left(\dfrac{2 + \sqrt{2}}{2}\right)}{2}} = \sqrt{\dfrac{2 + \sqrt{2}}{4}} = \dfrac{\sqrt{2 + \sqrt{2}}}{2}$

43. $\cos 15 = \cos \dfrac{30}{2} =$

$\sqrt{\dfrac{1 + \cos 30}{2}} = \sqrt{\dfrac{\left(1 + \dfrac{\sqrt{3}}{2}\right)}{2}} = \sqrt{\dfrac{\left(\dfrac{2 + \sqrt{3}}{2}\right)}{2}}$

$= \sqrt{\dfrac{2 + \sqrt{3}}{4}} = \dfrac{\sqrt{2 + \sqrt{3}}}{2}$

44.

t	X	Y
−2	3	−17
0	7	−1
5	17	39

45.

t	X	Y
−5	27	−18
−3	17	−12
4	−18	9

46. $X = 3t + 1 \rightarrow t = \dfrac{X - 1}{3}$

$Y = 3t - 7$

$Y = 3\left(\dfrac{X - 1}{3}\right) - 7$

$Y = X - 1 - 7$

$Y = X - 8$

47. $X = t - 1 \rightarrow t = X + 1$
    $Y = 2t + 7$
    $Y = 2(X + 1) + 7$
    $Y = 2X + 2 + 7$
    $Y = 2X + 9$

48. $X = 7\cos 37 = 5.59$
    $Y = 7\sin 37 = 4.21$
    $(5.59, 4.21)$

49. $r = \pm\sqrt{(5)^2 + (-4)^2}$
    $r = \sqrt{41}$
    $\tan\theta = \dfrac{-4}{5}$
    $\theta = \tan^{-1}\left(\dfrac{-4}{5}\right) = -38.7$
    $\left(\sqrt{41}, -38.7\right)$

50. $X = \left(5\sqrt{2}\right)(\cos 48.3) = 4.7$
    $Y = \left(5\sqrt{2}\right)(\sin 48.3) = 5.28$
    $(4.7, 5.28)$

Chapter 13   Mathematical induction and binomial theorem

▶ The mathematical induction is helpful in finding the statement is true for all natural numbers. If two conditions of the mathematical induction are met, then we have proven that the statement is true for all natural numbers.

The two conditions are

Condition I: For $n = 1$, the statement is true.

Condition II: Given k is true, the $k+1$ is true.

Example 1:
Apply the mathematical induction.

$$52 + 57 + 62 + 67 + \cdots + (47 + 5n) = \frac{99n + 5n^2}{2}$$

Condition I. $n = 1$

$47 + 5n = 47 + 5(1) = 52$ 　　　　We substitute $n = 1$ into $47 + 5n$ to obtain the first term.

$\dfrac{99n + 5n^2}{2} = \dfrac{99(1) + 5(1)^2}{2} = \dfrac{104}{2} = 52$ 　　We verify that the result is the same

as the previous one.

Condition I is met.

Condition II.

$52 + 57 + 62 + 67 + \cdots + (47 + 5k) = \dfrac{99k + 5k^2}{2}$ 　　We replace the n with k.

$52 + 57 + 62 + 67 + \cdots + (47 + 5k) + (47 + 5(k+1)) = \dfrac{99(k+1) + 5(k+1)^2}{2}$

We substitute $k + 1$.

$\underline{52 + 57 + 62 + 67 + \cdots + (47 + 5k)} + (5k + 52) = \dfrac{99(k+1) + 5(k+1)^2}{2}$

We replace the underlined part with what the original equation equals (notice that the underlined part is the same as the left side of the original equation). After the substitution, we will try to make both sides the same to determine if the condition II is true.

$$\frac{99k + 5k^2}{2} + (5k + 52) = \frac{99(k+1) + 5(k+1)^2}{2}$$

We determine whether the left side equals the right side.

$$\frac{99k + 5k^2}{2} + \frac{2(5k + 52)}{2} = \frac{99(k+1) + 5(k+1)^2}{2}$$

$$\frac{99k + 5k^2}{2} + \frac{10k + 104}{2} = \frac{99k + 99 + 5(k^2 + 2k + 1)}{2}$$

$$\frac{99k + 5k^2 + 10k + 104}{2} = \frac{99k + 99 + 5k^2 + 10k + 5}{2}$$

$$\frac{5k^2 + 109k + 104}{2} = \frac{5k^2 + 109k + 104}{2}$$

Both sides are equal.

Requirement for condition II is met.

Based on mathematical induction, the statement is true for all natural numbers.

Example 2:
Apply the mathematical induction.

$$3 + 9 + 27 + 81 + \cdots + 3^n = -\frac{3}{2}(1 - 3^n)$$

Condition I. $n = 1$

$$3^n = 3^1 = 3$$

We substitute $n = 1$ into 3^n to obtain the first term.

$$-\frac{3}{2}(1 - 3^n) = -\frac{3}{2}(1 - 3^1) = 3$$

We verify that the result is the same as the previous one.

Condition I is met.

Condition II.

$$3 + 9 + 27 + 81 + \cdots + 3^k = -\frac{3}{2}\left(1 - 3^k\right)$$
We replace the n with k.

$$3 + 9 + 27 + 81 + \cdots + 3^k + 3^{k+1} = -\frac{3}{2}\left(1 - 3^{k+1}\right)$$
We substitute $k + 1$.

$$\underline{3 + 9 + 27 + 81 + \cdots + 3^k} + 3^{k+1} = -\frac{3}{2}\left(1 - 3^{k+1}\right)$$

We replace the underlined part with what the original equation equals. After the substitution, we want to verify if both sides are the same.

$$-\frac{3}{2}\left(1 - 3^k\right) + 3^{k+1} = -\frac{3}{2}\left(1 - 3^{k+1}\right)$$

$$-\frac{3}{2} + \left(\frac{3}{2}\right)\left(3^k\right) + 3^{k+1} = -\frac{3}{2}\left(1 - 3^{k+1}\right)$$

$$-\frac{3}{2} + \left(\frac{3}{2}\right)\left(3^k\right) + \left(3^k\right)\left(3\right) = -\frac{3}{2}\left(1 - 3^{k+1}\right)$$

$$-\frac{3}{2} + \left(\frac{9}{2}\right)\left(3^k\right) = -\frac{3}{2}\left(1 - 3^{k+1}\right)$$

$$-\frac{3}{2}\left(1 - 3\left(3^k\right)\right) = -\frac{3}{2}\left(1 - 3^{k+1}\right)$$

$$-\frac{3}{2}\left(1 - 3^{k+1}\right) = -\frac{3}{2}\left(1 - 3^{k+1}\right)$$
Both sides are equal.

Requirement for condition II is met.

Based on mathematical induction, the statement is true for all natural numbers.

» The binomial theorem is a simpler way to expand the binomial to a large exponent power.

$$(aX + b)^n = \binom{n}{0}(b)^0(aX)^{n-0} + \binom{n}{1}(b)^1(aX)^{n-1} + \binom{n}{2}(b)^2(aX)^{n-2} + \cdots + \binom{n}{n}(b)^n(aX)^{n-n}$$

where $\dbinom{n}{r} = {}_nC_r = \dfrac{n!}{r!\,(n-r)}$

Example 3:

Apply the binomial theorem to expand.

$$(3X - 2)^8 = \qquad \text{We apply the binomial theorem formula to expand.}$$

$$(aX + b)^n = \binom{n}{0}(b)^0\,(aX)^{n-0} + \binom{n}{1}(b)^1\,(aX)^{n-1} + \binom{n}{2}(b)^2\,(aX)^{n-2} + \cdots + \binom{n}{n}(b)^n\,(aX)^{n-n}$$

$$\binom{8}{0}(-2)^0\,(3X)^{8-0} + \binom{8}{1}(-2)^1\,(3X)^{8-1} + \binom{8}{2}(-2)^2\,(3X)^{8-2} + \binom{8}{3}(-2)^3\,(3X)^{8-3}$$

$$+ \binom{8}{4}(-2)^4\,(3X)^{8-4} + \binom{8}{5}(-2)^5\,(3X)^{8-5} + \binom{8}{6}(-2)^6\,(3X)^{8-6} + \binom{8}{7}(-2)^7\,(3X)^{8-7}$$

$$+ \binom{8}{8}(-2)^8\,(3X)^{8-8}$$

$$\left(\frac{8!}{0!\,(8-0)}\right)(1)(3X)^8 + \left(\frac{8!}{1!\,(8-1)}\right)(-2)(3X)^7 + \left(\frac{8!}{2!\,(8-2)}\right)(4)(3X)^6 + \left(\frac{8!}{3!\,(8-3)}\right)(-8)(3X)^5$$

$$\left(\frac{8!}{4!\,(8-4)}\right)(16)(3X)^4 + \left(\frac{8!}{5!\,(8-5)}\right)(-32)(3X)^3 + \left(\frac{8!}{6!\,(8-6)}\right)(64)(3X)^2$$

$$\left(\frac{8!}{7!\,(8-7)}\right)(-128)(3X) + \left(\frac{8!}{8!\,(8-8)}\right)(256)(1)$$

$$6{,}561X^8 - 34{,}992X^7 + 81{,}648X^6 - 108{,}864X^5 + 90{,}720X^4 - 48{,}384X^3$$
$$+ 16{,}128X^2 - 3{,}072X + 256$$

Chapter 13   Mathematical induction and binomial theorem

1. Apply the mathematical induction. $5 + 16 + 27 + 38 + \cdots + (11n - 6) = \frac{1}{2}(11n^2 - n)$	2. Apply the mathematical induction. $2 + 7 + 12 + 17 + \cdots + (5n - 3) = \frac{1}{2}(5n^2 - n)$
3. Apply the mathematical induction. $3 + 9 + 15 + 21 + \cdots + (6n - 3) = 3n^2$	4. Apply the mathematical induction. $1 + 11 + 21 + 31 + \cdots + (10n - 9) = 5n^2 - 4n$
5. Apply the mathematical induction. $8 + 15 + 22 + 29 + \cdots + (7n + 1) = \frac{1}{2}(7n^2 + 9n)$	6. Apply the mathematical induction. $5 + 10 + 15 + 20 + \cdots + 5n = \frac{1}{2}(5n^2 + 5n)$
7. Apply the mathematical induction. $21 + 23 + 25 + 27 + \cdots + (2n + 19) = n^2 + 20n$	8. Apply the mathematical induction. $7 + 14 + 28 + 56 + \cdots + 7(2)^{n-1} = 7(2^n) - 7$
9. Apply the mathematical induction. $1 + 5 + 25 + 125 + \cdots + 5^{n-1} = \frac{1}{4}(5^n) - \frac{1}{4}$	10. Apply the mathematical induction. $2 + 6 + 18 + 54 + \cdots + 2(3)^{n-1} = 3^n - 1$

Chapter 13   Mathematical induction and binomial theorem

11. Apply the mathematical induction. $$3 + 6 + 12 + 24 + \cdots + 3(2)^{n-1} = 3(2^n) - 3$$	12. Apply the mathematical induction. $$7 + 12 + 17 + 22 + \cdots + (5n + 2)$$ $$= \frac{1}{2}(5n^2 + 9n)$$
13. Apply the mathematical induction. $$1 + 8 + 64 + 512 + \cdots + 8^{n-1} = \frac{1}{7}(8^n) - \frac{1}{7}$$	14. Apply the mathematical induction. $$4 + 6 + 8 + 10 + \cdots + (2n + 2) = n^2 + 3n$$
15. Apply the mathematical induction. $$2 + 8 + 32 + 128 + \cdots + 2^{2n-1} = \frac{2}{3}(4^n) - \frac{2}{3}$$	16. Apply the binomial theorem to expand. $$(X + 5)^7$$
17. Apply the binomial theorem to expand. $$(X - 1)^5$$	18. Apply the binomial theorem to expand. $$(2X + 1)^6$$
19. Apply the binomial theorem to expand. $$(3X - 4)^5$$	20. Apply the binomial theorem to expand. $$(5X - 1)^4$$

Chapter 13   Mathematical induction and binomial theorem

21. Apply the binomial theorem to expand. $(X-8)^4$	22. Apply the binomial theorem to expand. $(5X+2)^5$
23. Apply the binomial theorem to expand. $(4X-3)^5$	24. Apply the binomial theorem to expand. $(X+10)^4$
25. Apply the binomial theorem to expand. $(X-5)^6$	26. Apply the binomial theorem to expand. $(X+6)^5$
27. Apply the binomial theorem to expand. $(7X-1)^4$	28. Apply the binomial theorem to expand. $(3X+8)^5$
29. Apply the binomial theorem to expand. $(2X-4)^7$	30. Apply the binomial theorem to expand. $(5X-5)^5$

Chapter 13 Mathematical induction and binomial theorem
Answer key

1. Condition I.   n=1

   $$11n - 6 = 11(1) - 6 = 5 \text{ and } \frac{1}{2}\left(11 \cdot 1^2 - 1\right) = 5$$

   Requirement for condition I is met.

   Condition II.

   $$5 + 16 + 27 + 38 + \cdots + (11k - 6) = \frac{1}{2}\left(11k^2 - k\right)$$

   $$5 + 16 + 27 + 38 + \cdots + (11k - 6) + (11(k+1) - 6) = \frac{1}{2}\left(11(k+1)^2 - (k+1)\right)$$

   $$\underline{5 + 16 + 27 + 38 + \cdots + (11k - 6)} + (11k + 5) = \frac{1}{2}\left(11(k+1)^2 - (k+1)\right)$$

   $$\frac{1}{2}\left(11k^2 - k\right) + (11k + 5) = \frac{1}{2}\left(11(k+1)^2 - (k+1)\right)$$

   $$\frac{11}{2}k^2 - \frac{1}{2}k + 11k + 5 = \frac{1}{2}\left(11(k^2 + 2k + 1) - k - 1\right)$$

   $$\frac{11}{2}k^2 - \frac{1}{2}k + 11k + 5 = \frac{1}{2}\left(11k^2 + 22k + 11 - k - 1\right)$$

   $$\frac{11}{2}k^2 + \frac{21}{2}k + 5 = \frac{11}{2}k^2 + \frac{21}{2}k + 5$$

   Requirement for condition II is met.

   Based on mathematical induction, the statement is true for all natural numbers.

2. Condition I. n=1

$$5(1) - 3 = 2 \text{ and } \frac{1}{2}\left(5(1)^2 - 1\right) = \frac{1}{2}(5 - 1) = 2$$

Requirement for condition I is met.

Condition II.

$$2 + 7 + 12 + 17 + \cdots + (5k - 3) = \frac{1}{2}\left(5k^2 - k\right)$$

$$2 + 7 + 12 + 17 + \cdots + (5k - 3) + (5(k + 1) - 3) = \frac{1}{2}\left(5(k + 1)^2 - (k + 1)\right)$$

$$\underline{2 + 7 + 12 + 17 + \cdots + (5k - 3)} + (5k + 2) = \frac{1}{2}\left(5\left(k^2 + 2k + 1\right) - k - 1\right)$$

$$\frac{1}{2}\left(5k^2 - k\right) + (5k + 2) = \frac{1}{2}\left(5k^2 + 10k + 5 - k - 1\right)$$

$$\frac{5}{2}k^2 - \frac{1}{2}k + 5k + 2 = \frac{1}{2}\left(5k^2 + 9k + 4\right)$$

$$\frac{5}{2}k^2 + \frac{9}{2}k + 2 = \frac{5}{2}k^2 + \frac{9}{2}k + 2$$

Requirement for condition II is met.

Based on mathematical induction, the statement is true for all natural numbers.

3. Condition I. n=1

$$6(1) - 3 = 3 \text{ and } 3(1)^2 = 3$$

Requirement for condition I is met.

Condition II.

$$3 + 9 + 15 + 21 + \cdots + (6k - 3) = 3k^2$$

$$3 + 9 + 15 + 21 + \cdots + (6k - 3) + (6(k + 1) - 3) = 3(k + 1)^2$$

$$\underline{3 + 9 + 15 + 21 + \cdots + (6k - 3)} + (6k + 3) = 3\left(k^2 + 2k + 1\right)$$

$$3k^2 + (6k + 3) = 3k^2 + 6k + 3$$

$$3k^2 + 6k + 3 = 3k^2 + 6k + 3$$

Requirement for condition II is met.

Based on mathematical induction, the statement is true for all natural numbers.

4. Condition I. n=1

$10(1)-9=1$ and $5(1)^2-4(1)=1$

Requirement for condition I is met.

Condition II.

$1+11+21+31+\cdots+(10k-9)=5k^2-4k$

$1+11+21+31+\cdots+(10k-9)+(10(k+1)-9)=5(k+1)^2-4(k+1)$

$\underline{1+11+21+31+\cdots+(10k-9)+(10k+1)=5(k^2+2k+1)-4k-4}$

$5k^2-4k+(10k+1)=5k^2+10k+5-4k-4$

$5k^2+6k+1=5k^2+6k+1$

Requirement for condition II is met.

Based on mathematical induction, the statement is true for all natural numbers.

5. Condition I. n=1

$7(1)+1=8$ and $\frac{1}{2}(7\cdot1^2+9\cdot1)=\frac{1}{2}(7+9)=8$

Requirement for condition I is met.

Condition II.

$8+15+22+29+\cdots+(7k+1)=\frac{1}{2}(7k^2+9k)$

$8+15+22+29+\cdots+(7k+1)+(7(k+1)+1)=\frac{1}{2}(7(k+1)^2+9(k+1))$

$\underline{8+15+22+29+\cdots+(7k+1)+(7k+8)=\frac{1}{2}(7(k^2+2k+1)+9k+9)}$

$\frac{1}{2}(7k^2+9k)+(7k+8)=\frac{1}{2}(7k^2+14k+7+9k+9)$

$\frac{7}{2}k^2+\frac{9}{2}k+7k+8=\frac{1}{2}(7k^2+23k+16)$

$\frac{7}{2}k^2+\frac{23}{2}k+8=\frac{7}{2}k^2+\frac{23}{2}k+8$

Requirement for condition II is met.

Based on mathematical induction, the statement is true for all natural numbers.

6. Condition I. n=1

$$5(1)=5 \text{ and } \frac{1}{2}\left(5\cdot1^2+5\cdot1\right)=\frac{1}{2}(5+5)=5$$

Requirement for condition I is met.

Condition II.

$$5+10+15+20+\cdots+5k=\frac{1}{2}\left(5k^2+5k\right)$$

$$5+10+15+20+\cdots+5k+5(k+1)=\frac{1}{2}\left(5(k+1)^2+5(k+1)\right)$$

$$\underline{5+10+15+20+\cdots+5k}+5k+5=\frac{1}{2}\left(5\left(k^2+2k+1\right)+5k+5\right)$$

$$\frac{1}{2}\left(5k^2+5k\right)+5k+5=\frac{1}{2}\left(5k^2+10k+5+5k+5\right)$$

$$\frac{5}{2}k^2+\frac{5}{2}k+5k+5=\frac{1}{2}\left(5k^2+15k+10\right)$$

$$\frac{5}{2}k^2+\frac{15}{2}k+5=\frac{5}{2}k^2+\frac{15}{2}k+5$$

Requirement for condition II is met.

Based on mathematical induction, the statement is true for all natural numbers.

7. Condition I. n=1

$$2(1)+19=21 \text{ and } (1)^2+20(1)=21$$

Requirement for condition I is met.

Condition II.

$$21+23+25+27+\cdots+(2k+19)=k^2+20k$$

$$21+23+25+27+\cdots+(2k+19)+(2(k+1)+19)=(k+1)^2+20(k+1)$$

$$\underline{21+23+25+27+\cdots+(2k+19)}+(2k+21)=k^2+2k+1+20k+20$$

$$k^2+20k+(2k+21)=k^2+22k+21$$

$$k^2+22k+21=k^2+22k+21$$

Requirement for condition II is met.

Based on mathematical induction, the statement is true for all natural numbers.

8. Condition I. n=1

$$7(2)^{1-1} = 7(2)^0 = 7(1) = 7 \text{ and } 7(2^1) - 7 = 7(2) - 7 = 14 - 7 = 7$$

Requirement for condition I is met.

Condition II.

$$7 + 14 + 28 + 56 + \cdots + 7(2)^{k-1} = 7(2^k) - 7$$

$$7 + 14 + 28 + 56 + \cdots + 7(2)^{k-1} + 7(2)^{(k+1)-1} = 7(2^{(k+1)}) - 7$$

$$\underline{7 + 14 + 28 + 56 + \cdots + 7(2)^{k-1} + 7(2)^k = 7(2^{k+1}) - 7}$$

$$7(2^k) - 7 + 7(2)^k = 7(2^{k+1}) - 7$$

$$(2)(7)(2^k) - 7 = 7(2^{k+1}) - 7$$

$$7(2^{k+1}) - 7 = 7(2^{k+1}) - 7$$

Requirement for condition II is met.

Based on mathematical induction, the statement is true for all natural numbers.

9. Condition I. n=1

$$5^{(1)-1} = 5^0 = 1 \text{ and } \frac{1}{4}(5^{(1)}) - \frac{1}{4} = \frac{1}{4}(5) - \frac{1}{4} = \frac{5}{4} - \frac{1}{4} = \frac{4}{4} = 1$$

Requirement for condition I is met.

Condition II.

$$1 + 5 + 25 + 125 + \cdots + 5^{k-1} = \frac{1}{4}(5^k) - \frac{1}{4}$$

$$1 + 5 + 25 + 125 + \cdots + 5^{k-1} + 5^{(k+1)-1} = \frac{1}{4}(5^{(k+1)}) - \frac{1}{4}$$

$$\underline{1 + 5 + 25 + 125 + \cdots + 5^{k-1} + 5^k = \frac{1}{4}(5^{(k+1)}) - \frac{1}{4}}$$

$$\frac{1}{4}(5^k) - \frac{1}{4} + 5^k = \frac{1}{4}(5^{k+1}) - \frac{1}{4}$$

$$\frac{5}{4}(5^k) - \frac{1}{4} = \frac{1}{4}(5^k \cdot 5) - \frac{1}{4}$$

$$\frac{5}{4}(5^k) - \frac{1}{4} = \frac{5}{4}(5^k) - \frac{1}{4}$$

Requirement for condition II is met.

Based on mathematical induction, the statement is true for all natural numbers.

10. Condition I. $n=1$

$2(3)^{(1)-1} = 2(3)^0 = 2(1) = 2$ and $3^{(1)} - 1 = 3 - 1 = 2$

Requirement for condition I is met.

Condition II.

$2 + 6 + 18 + 54 + \cdots + 2(3)^{k-1} = 3^k - 1$

$2 + 6 + 18 + 54 + \cdots + 2(3)^{k-1} + 2(3)^{(k+1)-1} = 3^{(k+1)} - 1$

$\underline{2 + 6 + 18 + 54 + \cdots + 2(3)^{k-1} + 2(3)^k = 3^{k+1} - 1}$

$3^k - 1 + 2(3)^k = 3^{k+1} - 1$

$3(3^k) - 1 = 3^{k+1} - 1$

$3^{k+1} - 1 = 3^{k+1} - 1$

Requirement for condition II is met.

Based on mathematical induction, the statement is true for all natural numbers.

11. Condition I. $n=1$

$3(2)^{(1)-1} = 3(2)^0 = 3(1) = 3$ and $3(2^1) - 3 = 3(2) - 3 = 6 - 3 = 3$

Requirement for condition I is met.

Condition II.

$3 + 6 + 12 + 24 + \cdots + 3(2)^{k-1} = 3(2^k) - 3$

$3 + 6 + 12 + 24 + \cdots + 3(2)^{k-1} + 3(2)^{(k+1)-1} = 3(2^{(k+1)}) - 3$

$\underline{3 + 6 + 12 + 24 + \cdots + 3(2)^{k-1} + 3(2)^k = 3(2^{k+1}) - 3}$

$3(2^k) - 3 + 3(2)^k = 3(2^{k+1}) - 3$

$6(2^k) - 3 = 3(2^{k+1}) - 3$

$(3 \cdot 2)(2^k) - 3 = 3(2^{k+1}) - 3$

$3(2^{k+1}) - 3 = 3(2^{k+1}) - 3$

Requirement for condition II is met.

Based on mathematical induction, the statement is true for all natural numbers.

12. Condition I. n=1

$$5(1)+2=7 \text{ and } \frac{1}{2}\left(5(1)^2+9(1)\right)=\frac{1}{2}(5+9)=\frac{1}{2}(14)=7$$

Requirement for condition I is met.

Condition II.

$$7+12+17+22+\cdots+(5k+2)=\frac{1}{2}\left(5k^2+9k\right)$$

$$7+12+17+22+\cdots+(5k+2)+(5(k+1)+2)=\frac{1}{2}\left(5(k+1)^2+9(k+1)\right)$$

$$\underline{7+12+17+22+\cdots+(5k+2)}+(5k+7)=\frac{1}{2}\left(5(k^2+2k+1)+9k+9\right)$$

$$\frac{1}{2}\left(5k^2+9k\right)+(5k+7)=\frac{1}{2}\left(5(k^2+2k+1)+9k+9\right)$$

$$\frac{5}{2}k^2+\frac{9}{2}k+5k+7=\frac{1}{2}\left(5k^2+10k+5+9k+9\right)$$

$$\frac{5}{2}k^2+\frac{19}{2}k+7=\frac{1}{2}\left(5k^2+19k+14\right)$$

$$\frac{5}{2}k^2+\frac{19}{2}k+7=\frac{5}{2}k^2+\frac{19}{2}k+7$$

Requirement for condition II is met.

Based on mathematical induction, the statement is true for all natural numbers.

13. Condition I. n=1

$$8^{(1)-1}=8^0=1 \text{ and } \frac{1}{7}\left(8^{(1)}\right)-\frac{1}{7}=\frac{1}{7}(8)-\frac{1}{7}=\frac{8-1}{7}=\frac{7}{7}=1$$

Requirement for condition I is met.

Condition II.

$$1+8+64+512+\cdots+8^{k-1}=\frac{1}{7}\left(8^k\right)-\frac{1}{7}$$

$$1+8+64+512+\cdots+8^{k-1}+8^{(k+1)-1}=\frac{1}{7}\left(8^{(k+1)}\right)-\frac{1}{7}$$

$$\underline{1+8+64+512+\cdots+8^{k-1}}+8^k=\frac{1}{7}\left(8^{k+1}\right)-\frac{1}{7}$$

$$\frac{1}{7}\left(8^k\right)-\frac{1}{7}+8^k=\frac{1}{7}\left(8^{k+1}\right)-\frac{1}{7} \rightarrow \frac{8}{7}\left(8^k\right)-\frac{1}{7}=\frac{1}{7}\left(8^{k+1}\right)-\frac{1}{7}$$

$$\left(\frac{1}{7}\right)\left(8^{k+1}\right)-\frac{1}{7}=\frac{1}{7}\left(8^{k+1}\right)-\frac{1}{7}$$

Requirement for condition II is met.

Based on mathematical induction, the statement is true for all natural numbers.

14. Condition I. n=1

$$2(1)+2 = 4 \text{ and } (1)^2 + 3(1) = 1+3 = 4$$

Requirement for condition I is met.

Condition II.

$$4+6+8+10+\cdots+(2k+2) = k^2 + 3k$$

$$4+6+8+10+\cdots+(2k+2)+(2(k+1)+2) = (k+1)^2 + 3(k+1)$$

$$\underline{4+6+8+10+\cdots+(2k+2)+(2k+4) = (k^2+2k+1)+3k+3}$$

$$k^2 + 3k + (2k+4) = k^2 + 2k + 1 + 3k + 3$$

$$k^2 + 5k + 4 = k^2 + 5k + 4$$

Requirement for condition II is met.

Based on mathematical induction, the statement is true for all natural numbers.

15. Condition I. n=1

$$2^{2(1)-1} = 2^{2-1} = 2 \text{ and } \frac{2}{3}(4^1)-\frac{2}{3} = \frac{8}{3}-\frac{2}{3} = \frac{6}{3} = 2$$

Requirement for condition I is met.

Condition II.

$$2+8+32+128+\cdots+2^{2k-1} = \frac{2}{3}(4^k)-\frac{2}{3}$$

$$2+8+32+128+\cdots+2^{2k-1}+2^{2(k+1)-1} = \frac{2}{3}(4^{(k+1)})-\frac{2}{3}$$

$$\underline{2+8+32+128+\cdots+2^{2k-1}+2^{2k+1} = \frac{2}{3}(4^{k+1})-\frac{2}{3}}$$

$$\frac{2}{3}(4^k)-\frac{2}{3}+2^{2k+1} = \frac{2}{3}(4^{k+1})-\frac{2}{3}$$

$$\frac{2}{3}(4^k)-\frac{2}{3}+(2)(2^{2k}) = \frac{2}{3}(4^{k+1})-\frac{2}{3}$$

$$\frac{2}{3}(4^k)-\frac{2}{3}+(2)(4^k) = \frac{2}{3}(4^{k+1})-\frac{2}{3}$$

$$\frac{8}{3}(4^k)-\frac{2}{3} = \frac{2}{3}(4^{k+1})-\frac{2}{3}$$

$$\frac{2}{3}(4)(4^k)-\frac{2}{3} = \frac{2}{3}(4^{k+1})-\frac{2}{3}$$

$$\frac{2}{3}(4^{k+1})-\frac{2}{3} = \frac{2}{3}(4^{k+1})-\frac{2}{3}$$

Requirement for condition II is met

Based on mathematical induction, the statement is true for all natural numbers.

16. $\binom{7}{0}(5)^0(X)^{7-0}+\binom{7}{1}(5)^1(X)^{7-1}+\binom{7}{2}(5)^2(X)^{7-2}+\binom{7}{3}(5)^3(X)^{7-3}+\binom{7}{4}(5)^4(X)^{7-4}$

$+\binom{7}{5}(5)^5(X)^{7-5}+\binom{7}{6}(5)^6(X)^{7-6}+\binom{7}{7}(5)^7(X)^{7-7}$

$=X^7+35X^6+525X^5+4{,}375X^4+21{,}875X^3+65{,}625X^2+109{,}375X+78{,}125$

17. $\binom{5}{0}(-1)^0(X)^{5-0}+\binom{5}{1}(-1)^1(X)^{5-1}+\binom{5}{2}(-1)^2(X)^{5-2}+\binom{5}{3}(-1)^3(X)^{5-3}$

$+\binom{5}{4}(-1)^4(X)^{5-4}+\binom{5}{5}(-1)^5(X)^{5-5}$

$=X^5-5X^4+10X^3-10X^2+5X-1$

18. $\binom{6}{0}(1)^0(2X)^{6-0}+\binom{6}{1}(1)^1(2X)^{6-1}+\binom{6}{2}(1)^2(2X)^{6-2}+\binom{6}{3}(1)^3(2X)^{6-3}+\binom{6}{4}(1)^4(2X)^{6-4}$

$+\binom{6}{5}(1)^5(2X)^{6-5}+\binom{6}{6}(1)^6(2X)^{6-6}$

$=64X^6+192X^5+240X^4+160X^3+60X^2+12X+1$

19. $\binom{5}{0}(-4)^0(3X)^{5-0}+\binom{5}{1}(-4)^1(3X)^{5-1}+\binom{5}{2}(-4)^2(3X)^{5-2}+\binom{5}{3}(-4)^3(3X)^{5-3}$

$+\binom{5}{4}(-4)^4(3X)^{5-4}+\binom{5}{5}(-4)^5(3X)^{5-5}$

$=243X^5-1{,}620X^4+4{,}320X^3-5{,}760X^2+3{,}840X-1{,}024$

20. $\binom{4}{0}(-1)^0(5X)^{4-0}+\binom{4}{1}(-1)^1(5X)^{4-1}+\binom{4}{2}(-2)^2(5X)^{4-2}+\binom{4}{3}(-2)^3(5X)^{4-3}$

$+\binom{4}{4}(-2)^4(5X)^{4-4}$

$=625X^4-500X^3+150X^2-20X+1$

21. $\binom{4}{0}(-8)^0(X)^{4-0}+\binom{4}{1}(-8)^1(X)^{4-1}+\binom{4}{2}(-8)^2(X)^{4-2}+\binom{4}{3}(-8)^3(X)^{4-3}$

$+\binom{4}{4}(-8)^4(X)^{4-4}$

$=X^4-32X^3+384X^2-2{,}048X+4{,}096$

22. $\binom{5}{0}(2)^0(5X)^{5-0}+\binom{5}{1}(2)^1(5X)^{5-1}+\binom{5}{2}(2)^2(5X)^{5-2}+\binom{5}{3}(2)^3(5X)^{5-3}$

$+\binom{5}{4}(2)^4(5X)^{5-4}+\binom{5}{5}(2)^5(5X)^{5-5}$

$=3{,}125X^5+6{,}250X^4+5{,}000X^3+2{,}000X^2+400X+32$

23. $\binom{5}{0}(-3)^0(4X)^{5-0}+\binom{5}{1}(-3)^1(4X)^{5-1}+\binom{5}{2}(-3)^2(4X)^{5-2}+\binom{5}{3}(-3)^3(4X)^{5-3}$

$+\binom{5}{4}(-3)^4(4X)^{5-4}+\binom{5}{5}(-3)^5(4X)^{5-5}$

$=1{,}024X^5-3{,}840X^4+5{,}760X^3-4{,}320X^2+1{,}620X-243$

24. $\binom{4}{0}(10)^0(X)^{4-0}+\binom{4}{1}(10)^1(X)^{4-1}+\binom{4}{2}(10)^2(X)^{4-2}+\binom{4}{3}(10)^3(X)^{4-3}$

$\binom{4}{4}(10)^4(X)^{4-4}$

$=X^4+40X^3+600X^2+4{,}000X+10{,}000$

25. $\binom{6}{0}(-5)^0(X)^{6-0}+\binom{6}{1}(-5)^1(X)^{6-1}+\binom{6}{2}(-5)^2(X)^{6-2}+\binom{6}{3}(-5)^3(X)^{6-3}$

$\binom{6}{4}(-5)^4(X)^{6-4}+\binom{6}{5}(-5)^5(X)^{6-5}+\binom{6}{6}(-5)^6(X)^{6-6}$

$=X^6-30X^5+375X^4-2{,}500X^3+9{,}375X^2-18{,}750X+15{,}625$

26. $\binom{5}{0}(6)^0(X)^{5-0}+\binom{5}{1}(6)^1(X)^{5-1}+\binom{5}{2}(6)^2(X)^{5-2}+\binom{5}{3}(6)^3(X)^{5-3}$

$+\binom{5}{4}(6)^4(X)^{5-4}+\binom{5}{5}(6)^5(X)^{5-5}$

$=X^5+30X^4+360X^3+2{,}160X^2+6{,}480X+7{,}776$

27. $\binom{4}{0}(-1)^0(7X)^{4-0}+\binom{4}{1}(-1)^1(7X)^{4-1}+\binom{4}{2}(-1)^2(7X)^{4-2}+\binom{4}{3}(-1)^3(7X)^{4-3}$

$\binom{4}{4}(-1)^4(7X)^{4-4}$

$=2{,}401X^4-1{,}372X^3+294X^2-28X+1$

28. $\binom{5}{0}(8)^0(3X)^{5-0} + \binom{5}{1}(8)^1(3X)^{5-1} + \binom{5}{2}(8)^2(3X)^{5-2} + \binom{5}{3}(8)^3(3X)^{5-3}$

$+ \binom{5}{4}(8)^4(3X)^{5-4} + \binom{5}{5}(8)^5(3X)^{5-5}$

$= 243X^5 + 3{,}240X^4 + 17{,}280X^3 + 46{,}080X^2 + 61{,}440X + 32{,}768$

29. $\binom{7}{0}(-4)^0(2X)^{7-0} + \binom{7}{1}(-4)^1(2X)^{7-1} + \binom{7}{2}(-4)^2(2X)^{7-2} + \binom{7}{3}(-4)^3(2X)^{7-3}$

$+ \binom{7}{4}(-4)^4(2X)^{7-4} + \binom{7}{5}(-4)^5(2X)^{7-5} + \binom{7}{6}(-4)^6(2X)^{7-6} + \binom{7}{7}(-4)^7(2X)^{7-7}$

$= 128X^7 - 1{,}792X^6 + 10{,}752X^5 - 35{,}840X^4 + 71{,}680X^3 - 86{,}016X^2 + 57{,}344X - 16{,}384$

30. $\binom{5}{0}(-5)^0(5X)^{5-0} + \binom{5}{1}(-5)^1(5X)^{5-1} + \binom{5}{2}(-5)^2(5X)^{5-2} + \binom{5}{3}(-5)^3(5X)^{5-3}$

$+ \binom{5}{4}(-5)^4(5X)^{5-4} + \binom{5}{5}(-5)^5(5X)^{5-5}$

$= 3{,}125X^5 - 15{,}625X^4 + 31{,}250X^3 - 31{,}250X^2 + 15{,}625X - 3{,}125$

Chapter 14   Limits

▶ The limit exists if the left limit and the right limit are equal. We find the value of left limit by gradually approaching the value from the left. We find the right limit by gradually approaching the value from the right.

Left limit: $\lim_{X \to c^-} f(X) = value$

Right limit: $\lim_{X \to c^+} f(X) = value$

$$\lim_{X \to c^-} f(X) = \lim_{X \to c} f(X) = \lim_{X \to c^+} f(X)$$

Limit exists when the left limit equals the right limit.

Example 1:
Evaluate the limit by completing the table.

$$\lim_{X \to 2}(5X + 2) =$$

X	1.8	1.9	1.99	2	2.01	2.1	2.2
Y				■			

$5(1.8) + 2 = 11$
$5(1.9) + 2 = 11.5$
$5(1.99) + 2 = 11.95$

$5(2.2) + 2 = 13$
$5(2.1) + 2 = 12.5$
$5(2.01) + 2 = 12.05$

We substitute the corresponding x-coordinate into the equation to evaluate the y-coordinate.

X	1.8	1.9	1.99	2	2.01	2.1	2.2
Y	11	11.5	11.95	■	12.05	12.5	13

As the limit approaches from the left, the limit approaches toward 12. The left limit is 12.

As the limit approaches from the right, the limit approaches toward 12. The right limit is 12.

We write the corresponding y-coordinates in the chart. We find the left limit and the right limit. Because the left limit and right limit both equal 12, the limit as X approaches toward 2 is 12.

$$\lim_{X \to 2}(5X + 2) = 12$$

Example 2:

Evaluate the limit by completing
the table.

$$\lim_{X \to 5} \frac{X^2 + 5X - 1}{X - 5} =$$

X	4.8	4.9	4.99	5	5.01	5.1	5.2
Y				■			

$f(4.8) = -230.2$

$f(4.9) = -475.1$

$f(4.99) = -4,885$

We substitute the x-coordinates into the equation to find the corresponding y-coordinates.

$f(5.2) = 260.2$

$f(5.1) = 505.1$

$f(5.01) = 4,915$

X	4.8	4.9	4.99	5	5.01	5.1	5.2
Y	-230.2	-475.1	$-4,885$	■	4,915	505.1	260.2

$$\lim_{X \to 5^-} \frac{X^2 + 5X - 1}{X - 5} = -4.900$$

The left limit appears to approach the value $-4,900$.

$$\lim_{X \to 5^+} \frac{X^2 + 5X - 1}{X - 5} = 4,900$$

The right limit appears to approach the value 4,900.

$$\lim_{X \to 5^-} \frac{X^2 + 5X - 1}{X - 5} \neq \lim_{X \to 5^+} \frac{X^2 + 5X - 1}{X - 5}$$

The left limit does not equal the right limit. Therefore, the limit does not exist as the limit approaches $X = 5$.

» We can apply the previous table method to evaluate the limit. We can also apply the limit rules to evaluate the limits. We substitute the approaching number into the equation to evaluate the limit.

$$\lim_{X \to c}(A + B) = \lim_{X \to c} A + \lim_{X \to c} B$$

$$\lim_{X \to c}(A - B) = \lim_{X \to c} A - \lim_{X \to c} B$$

$$\lim_{X \to c}(aX) = a \lim_{X \to c} X \qquad \text{("} a \text{" is a constant number.)}$$

Example 3:
Evaluate the limit.

$$\lim_{X \to 2}\left(5X^3 + 2X^2 - 8X + 1\right) =$$

$$\lim_{X \to 2}\left(5X^3\right) + \lim_{X \to 2}\left(2X^2\right) - \lim_{X \to 2}\left(8X\right) + \lim_{X \to 2} 1 =$$

We apply the limit rules:

$$\lim_{X \to c}(A + B) = \lim_{X \to c} A + \lim_{X \to c} B$$
$$\lim_{X \to c}(A - B) = \lim_{X \to c} A - \lim_{X \to c} B .$$

$$5 \lim_{X \to 2} X^3 + 2 \lim_{X \to 2} X^2 - 8 \lim_{X \to 2} X + \lim_{X \to 2} 1 =$$

We apply the limit rule:

$$\lim_{X \to c}(aX) = a \lim_{X \to c} X .$$

$$5(2)^3 + 2(2)^2 - 8(2) + 1 = 33$$

We substitute the approaching number into X and evaluate.

$$\lim_{X \to 2}\left(5X^3 + 2X^2 - 8X + 1\right) = 33$$

Example 4:
Evaluate the limit.

$$\lim_{X \to (-5)}\left(\frac{2X^2 + 5X - 1}{3X - 2}\right) =$$

$$\frac{\lim_{X \to (-5)}\left(2X^2\right) + \lim_{X \to (-5)}(5X) - \lim_{X \to (-5)} 1}{\lim_{X \to (-5)}(3X) - \lim_{X \to (-5)} 2} =$$

We apply the limit rules:

$$\lim_{X \to c}(A + B) = \lim_{X \to c} A + \lim_{X \to c} B$$
$$\lim_{X \to c}(A - B) = \lim_{X \to c} A - \lim_{X \to c} B .$$

$$\frac{2 \lim_{X \to (-5)} X^2 + 5 \lim_{X \to (-5)} X - \lim_{X \to (-5)} 1}{3 \lim_{X \to (-5)} X - \lim_{X \to (-5)} 2} =$$

We apply the limit rule

$$\lim_{X \to c}(aX) = a \lim_{X \to c} X .$$

$$\frac{2(-5)^2 + 5(-5) - 1}{3(-5) - 2} =$$

We substitute the

approaching number into the equation and evaluate.

$$\frac{24}{-17} = -\frac{24}{17}$$

$$\lim_{X \to (-5)} \left(\frac{2X^2 + 5X - 1}{3X - 2} \right) = -\frac{24}{17}$$

» We may not be able to substitute the approaching number into X because the equation may be undefined. In this case, we can either apply the table method (the method we studied in the beginning) or simplify the equation. In the next example, we will simplify the equation.

Example 5:
Evaluate the limit.

$$\lim_{X \to 7} \frac{X^2 - 2X - 35}{X^2 - 6X - 7} =$$

$$\frac{\lim_{X \to 7} X^2 - 2 \lim_{X \to 7} X - \lim_{X \to 7} 35}{\lim_{X \to 7} X^2 - 6 \lim_{X \to 7} X - \lim_{X \to 7} 7} =$$

$$\frac{(7)^2 - 2(7) - 35}{(7)^2 - 6(7) - 7} = \frac{0}{0} = undefined$$

The limit is undefined. This does not

mean we cannot find the limit. We simplify the equation and evaluate the limit.

$$\lim_{X \to 7} \frac{X^2 - 2X - 35}{X^2 - 6X - 7} =$$

$$\lim_{X \to 7} \frac{(X-7)(X+5)}{(X-7)(X+1)} =$$

We factor the numerator and the denominator.

$$\lim_{X \to 7} \frac{X+5}{X+1} =$$

We cancel the common factors in the numerator and in the denominator.

$$\frac{\lim_{X \to 7} X + \lim_{X \to 7} 5}{\lim_{X \to 7} X + \lim_{X \to 7} 1} =$$

We apply the limit rules.

$$\frac{7+5}{7+1} = \frac{12}{8} = \frac{3}{2}$$

We evaluate the limit.

$$\lim_{X \to 7} \frac{X^2 - 2X - 35}{X^2 - 6X - 7} = \frac{3}{2}$$

» When we have X approaching infinity and we have a fraction equation, we can evaluate the limit by dividing each term in the equation by the largest degree. Then, we substitute approaching value (the infinity) in the X and evaluate the limit. A number over an infinity is equal to 0 $\left(\frac{5}{\infty} = 0 \right)$.

Example 6:
Evaluate the limit.

$$\lim_{X \to \infty} \left(\frac{5X^3 + 2X^2 - 1}{8X^3 - X + 8} \right) =$$

We have a limit with X approaching infinity. The largest degree is X^3.

$$\lim_{X \to \infty} \frac{\dfrac{5X^3}{X^3} + \dfrac{2X^2}{X^3} - \dfrac{1}{X^3}}{\dfrac{8X^3}{X^3} - \dfrac{X}{X^3} + \dfrac{8}{X^3}} =$$

We divide each term by X^3.

$$\lim_{X \to \infty} \frac{5 + \dfrac{2}{X} - \dfrac{1}{X^3}}{8 - \dfrac{1}{X^2} + \dfrac{8}{X^3}} =$$

$$\frac{\lim\limits_{X \to \infty} 5 + \lim\limits_{X \to \infty} \dfrac{2}{X} - \lim\limits_{X \to \infty} \dfrac{1}{X^3}}{\lim\limits_{X \to \infty} 8 - \lim\limits_{X \to \infty} \dfrac{1}{X^2} + \lim\limits_{X \to \infty} \dfrac{8}{X^3}} =$$

We apply the limit rules.

$$\frac{5+0-0}{8-0+0} = \frac{5}{8}$$

Remember that $\dfrac{number}{\infty} = 0$.

$$\lim\limits_{X \to \infty} \left(\frac{5X^3 + 2X^2 - 1}{8X^3 - X + 8} \right) = \frac{5}{8}$$

» We can determine and verify whether the graph is continuous at a given point. If the following three requirements are met, then the graph is continuous at the given point:

a). $f(c) = value$

The function of the equation at the given point has a value.

b). $\lim\limits_{X \to c^-} f(X) = \lim\limits_{X \to c} f(X) = \lim\limits_{X \to c^+} f(X)$

The limit exists (the left limit equals the right limit).

c). $f(c) = \lim\limits_{X \to c} f(X)$

The value of $f(c)$ is same as the value of limit $\lim\limits_{X \to c} f(X)$.

Example 7:
Determine whether $f(X)$ is continuous at $X = 3$.

$$f(X) = \frac{X^2 + 3X + 1}{X + 2}$$

a). $f(3) = \dfrac{(3)^2 + 3(3) + 1}{(3) + 2} = \dfrac{19}{5}$

The first requirement is met. The function has a value at $X = 3$.

b). $\lim\limits_{X \to 3^-} \dfrac{X^2 + 3X + 1}{X + 2} = \dfrac{19}{5} = \lim\limits_{X \to 3^+} \dfrac{X^2 + 3X + 1}{X + 2}$

The left limit equals the right limit. The limit exists.

$$\lim_{X \to 3} \frac{X^2 + 3X + 1}{X + 2} = \frac{19}{5}$$

The second requirement is met. The limit exists at $X = 3$.

c). $f(3) = \dfrac{19}{5} = \lim_{X \to 3} \dfrac{X^2 + 3X + 1}{X + 2}$

$f(3) = \lim_{X \to 3} f(X)$

The third requirement is met. The value of function equals the value of limit at $X = 3$.

Because all the three requirements are met, the graph is continuous at $X = 3$.

Chapter 14   Limits

1. Evaluate the limit by completing the table. $$\lim_{X \to 2}(-2X+3)=$$	2. Evaluate the limit by completing the table. $$\lim_{X \to 6}\left(\frac{8}{X-6}\right)=$$

X	1.8	1.9	1.99	2	2.01	2.1	2.2
Y				■			

X	5.8	5.9	5.99	6	6.01	6.1	6.2
Y				■			

3. Evaluate the limit by completing the table. $$\lim_{X \to 8}\left(\frac{X^2-121}{X+11}\right)=$$	4. Evaluate the limit by completing the table. $$\lim_{X \to 1}(12X+8)=$$

X	0.8	0.9	0.99	1	1.01	1.1	1.2
Y				■			

X	7.8	7.9	7.99	8	8.01	8.1	8.2
Y				■			

5. Evaluate the limit by completing the table. $$\lim_{X \to 2}\left(\frac{X+21}{X-2}\right)=$$	6. Evaluate the limit by completing the table. $$\lim_{X \to 4}(-3X-11)=$$

X	3.8	3.9	3.99	4	4.01	4.1	4.2
Y				■			

X	1.8	1.9	1.99	2	2.01	2.1	2.2
Y				■			

7. Evaluate the limit by completing the table. $$\lim_{X \to 6}\left(\frac{X^2+4X-1}{X+2}\right)=$$	8. Evaluate the limit by completing the table. $$\lim_{X \to 4}\left(\frac{X^2+14}{X-4}\right)=$$

X	5.8	5.9	5.99	6	6.01	6.1	6.2
Y				■			

X	3.8	3.9	3.99	4	4.01	4.1	4.2
Y				■			

9. Evaluate the limit by completing the table. $$\lim_{X \to 5}(3X^2+2X-1)=$$	10. Evaluate the limit by completing the table. $$\lim_{X \to 2}(4X^2-4X+1)=$$

X	4.8	4.9	4.99	5	5.01	5.1	5.2
Y				■			

X	1.8	1.9	1.99	2	2.01	2.1	2.2
Y				■			

Chapter 14   Limits

11. Evaluate the limit. $$\lim_{X \to 7}(3X + 4)=$$	12. Evaluate the limit. $$\lim_{X \to (-2)}\left(2X^2 - 4X + 1\right)=$$
13. Evaluate the limit. $$\lim_{X \to 12}\left(X^2 - X + 7\right)=$$	14. Evaluate the limit. $$\lim_{X \to (-3)}\left(3X^3 - 2X^2 + X + 1\right)=$$
15. Evaluate the limit. $$\lim_{X \to 1}\left(5X^3 - 2X^2 + 3X - 3\right)=$$	16. Evaluate the limit. $$\lim_{X \to 4}\left(X^4 - 3X^3 + X + 11\right)=$$
17. Evaluate the limit. $$\lim_{X \to 3}\left(X^5 - 4X^4 + 3X^2 + 2X - 1\right)=$$	18. Evaluate the limit. $$\lim_{X \to (-2)}\left(\frac{X^2 + 2X - 1}{3X + 1}\right)$$
19. Evaluate the limit. $$\lim_{X \to 1}\left(\frac{X^2 + 4}{3X + 1}\right)=$$	20. Evaluate the limit. $$\lim_{X \to (-5)}\left(\frac{X^2 + 4X - 1}{5X - 1}\right)=$$

Chapter 14   Limits

21. Evaluate the limit. $$\lim_{X \to (-15)}\left(\frac{X^2 - 225}{X + 15}\right) =$$	22. Evaluate the limit. $$\lim_{X \to 20}\left(\frac{X - 20}{X^2 - 400}\right) =$$
23. Evaluate the limit. $$\lim_{X \to (-17)}\left(\frac{X^2 + 14X - 51}{X + 17}\right) =$$	24. Evaluate the limit. $$\lim_{X \to (-11)}\left(\frac{X^2 + 8X - 33}{X^2 + 12X + 11}\right) =$$
25. Evaluate the limit. $$\lim_{X \to 5}\left(\frac{X^2 + 10X - 75}{X^2 + 3X - 40}\right) =$$	26. Evaluate the limit. $$\lim_{X \to \infty}\left(\frac{7X^2 + 2X - 4}{X^2 + 4X - 1}\right) =$$
27. Evaluate the limit. $$\lim_{X \to \infty}\left(\frac{3X^3 - 4X^2 - 4}{4X^4 - 3X^2 - 1}\right) =$$	28. Evaluate the limit. $$\lim_{X \to \infty}\left(\frac{5X^4 - 3X^2 + 7}{3X^4 + 2X^2 - 1}\right) =$$
29. Evaluate the limit. $$\lim_{X \to \infty}\left(\frac{3X^2 - 2X + 1}{12X^2 + 5X - 1}\right) =$$	30. Evaluate the limit. $$\lim_{X \to \infty}\left(\frac{8X^3 - 2X^2 + X + 1}{7X^3 - 4X - 1}\right) =$$

Chapter 14   Limits

31. Determine whether $f(X)$ is continuous at $X = 7$ $f(X) = 2X - 5$	32. Determine whether $f(X)$ is continuous at $X = 5$ $f(X) = \dfrac{3}{X-5}$
33. Determine whether $f(X)$ is continuous at $X = 5$ $f(X) = X^2 - 3$	34. Determine whether $f(X)$ is continuous at $X = 5$ $f(X) = \dfrac{5}{X-1}$
35. Determine whether $f(X)$ is continuous at $X = -2$ $f(X) = \dfrac{2X+5}{X^2+5X+6}$	36. Determine whether $f(X)$ is continuous at $X = 7$ $f(X) = \dfrac{X-7}{X^2+3X+2}$
37. Determine whether $f(X)$ is continuous at $X = 11$ $f(X) = \dfrac{2X-1}{X^2-15X+44}$	

Chapter 14 Limits
Answer key

1.

X	1.8	1.9	1.99	2	2.01	2.1	2.2
Y	−0.6	−0.8	−0.98	■	−1.02	−1.2	−1.4

$$\lim_{X \to 2^-}(-2X+3) = -1 = \lim_{X \to 2^+}(-2X+3)$$

$$\lim_{X \to 2}(-2X+3) = -1$$

2.

X	5.8	5.9	5.99	6	6.01	6.1	6.2
Y	−40	−80	−800	■	800	80	40

$$\lim_{X \to 6^-}\left(\frac{8}{X-6}\right) \neq \lim_{X \to 6^+}\left(\frac{8}{X-6}\right)$$

The limit does not exist.

3.

X	7.8	7.9	7.99	8	8.01	8.1	8.2
Y	−3.2	−3.1	−3.01	■	−2.99	−2.9	−2.8

$$\lim_{X \to 8^-}\left(\frac{X^2-121}{X+11}\right) = -3 = \lim_{X \to 8^+}\left(\frac{X^2-121}{X+11}\right)$$

$$\lim_{X \to 8}\left(\frac{X^2-121}{X+11}\right) = -3$$

4.

X	0.8	0.9	0.99	1	1.01	1.1	1.2
Y	17.6	18.8	19.88	■	20.12	21.2	22.4

$$\lim_{X \to 1^-}(12X+8) = 20 = \lim_{X \to 1^+}(12X+8)$$

$$\lim_{X \to 1}(12X+8) = 20$$

5.

X	1.8	1.9	1.99	2	2.01	2.1	2.2
Y	−114	−229	−2,299	■	2,301	231	116

$$\lim_{X \to 2^-}\left(\frac{X+21}{X-2}\right) \neq \lim_{X \to 2^+}\left(\frac{X+21}{X-2}\right)$$

The limit does not exist.

6.

X	3.8	3.9	3.99	4	4.01	4.1	4.2
Y	-22.4	-22.7	-22.97	■	-23.03	-23.3	-23.6

$$\lim_{X \to 4^-}\left(-3X-11\right)=-23=\lim_{X \to 4^+}\left(-3X-11\right)$$

$$\lim_{X \to 4}\left(-3X-11\right)=-23$$

7.

X	5.8	5.9	5.99	6	6.01	6.1	6.2
Y	7.159	7.2671	7.3642	■	7.3858	7.4827	7.5902

$$\lim_{X \to 6^-}\left(\frac{X^2+4X-1}{X+2}\right)=7.375=\lim_{X \to 6^+}\left(\frac{X^2+4X-1}{X+2}\right)$$

$$\lim_{X \to 6}\left(\frac{X^2+4X-1}{X+2}\right)=7.375$$

8.

X	3.8	3.9	3.99	4	4.01	4.1	4.2
Y	-142.2	-292.1	$-2,992$	■	$3,008$	308.1	158.2

$$\lim_{X \to 4^-}\left(\frac{X^2+14}{X-4}\right) \neq \lim_{X \to 4^+}\left(\frac{X^2+14}{X-4}\right)$$

The limit does not exist.

9.

X	4.8	4.9	4.99	5	5.01	5.1	5.2
Y	77.72	80.83	83.68	■	84.32	87.23	90.52

$$\lim_{X \to 5^-}\left(3X^2+2X-1\right)=84=\lim_{X \to 5^+}\left(3X^2+2X-1\right)$$

$$\lim_{X \to 5}\left(3X^2+2X-1\right)=84$$

10.

X	1.8	1.9	1.99	2	2.01	2.1	2.2
Y	6.76	7.84	8.88	■	9.12	10.24	11.56

$$\lim_{X \to 2^-}\left(4X^2-4X+1\right)=9=\lim_{X \to 2^+}\left(4X^2-4X+1\right)$$

$$\lim_{X \to 2}\left(4X^2-4X+1\right)=9$$

11. $\lim_{X \to 7}\left(3X+4\right)=\lim_{X \to 7}3X+\lim_{X \to 7}4=3\lim_{X \to 7}X+\lim_{X \to 7}4=3(7)+4=25$

12. $\lim\limits_{X\to(-2)}\left(2X^2-4X+1\right)=\lim\limits_{X\to(-2)}2X^2-\lim\limits_{X\to(-2)}4X+\lim\limits_{X\to(-2)}1$

$=2\lim\limits_{X\to(-2)}X^2-4\lim\limits_{X\to(-2)}X+\lim\limits_{X\to(-2)}1=2(-2)^2-4(-2)+1=17$

13. $\lim\limits_{X\to12}\left(X^2-X+7\right)=\lim\limits_{X\to12}X^2-\lim\limits_{X\to12}X+\lim\limits_{X\to12}7=(12)^2-12+7=144-12+7=139$

14. $\lim\limits_{X\to(-3)}\left(3X^3-2X^2+X+1\right)=\lim\limits_{X\to(-3)}3X^3-\lim\limits_{X\to(-3)}2X^2+\lim\limits_{X\to(-3)}X+\lim\limits_{X\to(-3)}1$

$=3\lim\limits_{X\to(-3)}X^3-2\lim\limits_{X\to(-3)}X^2+\lim\limits_{X\to(-3)}X+\lim\limits_{X\to(-3)}1=3(-3)^3-2(-3)^2+(-3)+1=-101$

15. $\lim\limits_{X\to1}\left(5X^3-2X^2+3X-3\right)=\lim\limits_{X\to1}5X^3-\lim\limits_{X\to1}2X^2+\lim\limits_{X\to1}3X-\lim\limits_{X\to1}3$

$=5\lim\limits_{X\to1}X^3-2\lim\limits_{X\to1}X^2+3\lim\limits_{X\to1}X-\lim\limits_{X\to1}3=5(1)^3-2(1)^2+3(1)-3=3$

16. $\lim\limits_{X\to4}\left(X^4-3X^3+X+11\right)=\lim\limits_{X\to4}X^4-\lim\limits_{X\to4}3X^3+\lim\limits_{X\to4}X+\lim\limits_{X\to4}11$

$=\lim\limits_{X\to4}X^4-3\lim\limits_{X\to4}X^3+\lim\limits_{X\to4}X+\lim\limits_{X\to4}11=(4)^4-3(4)^3+4+11=79$

17. $\lim\limits_{X\to3}\left(X^5-4X^4+3X^2+2X-1\right)=\lim\limits_{X\to3}X^5-\lim\limits_{X\to3}4X^4+\lim\limits_{X\to3}3X^2+\lim\limits_{X\to3}2X-\lim\limits_{X\to3}1$

$\lim\limits_{X\to3}X^5-4\lim\limits_{X\to3}X^4+3\lim\limits_{X\to3}X^2+2\lim\limits_{X\to3}X-\lim\limits_{X\to3}1=(3)^5-4(3)^4+3(3)^2+2(3)-1=-49$

18. $\lim\limits_{X\to(-2)}\left(\dfrac{X^2+2X-1}{3X+1}\right)=\dfrac{\lim\limits_{X\to(-2)}X^2+\lim\limits_{X\to(-2)}2X-\lim\limits_{X\to(-2)}1}{\lim\limits_{X\to(-2)}3X+\lim\limits_{X\to(-2)}1}$

$=\dfrac{\lim\limits_{X\to(-2)}X^2+2\lim\limits_{X\to(-2)}X-\lim\limits_{X\to(-2)}1}{3\lim\limits_{X\to(-2)}X+\lim\limits_{X\to(-2)}1}=\dfrac{(-2)^2+2(-2)-1}{3(-2)+1}=\dfrac{1}{5}$

19. $\lim\limits_{X\to1}\left(\dfrac{X^2+4}{3X+1}\right)=\dfrac{\lim\limits_{X\to1}X^2+\lim\limits_{X\to1}4}{\lim\limits_{X\to1}3X+\lim\limits_{X\to1}1}=\dfrac{\lim\limits_{X\to1}X^2+\lim\limits_{X\to1}4}{3\lim\limits_{X\to1}X+\lim\limits_{X\to1}1}=\dfrac{(1)^2+4}{3(1)+1}=\dfrac{5}{4}$

20. $\lim\limits_{X\to(-5)}\left(\dfrac{X^2+4X-1}{5X-1}\right)=\dfrac{\lim\limits_{X\to(-5)}X^2+\lim\limits_{X\to(-5)}4X-\lim\limits_{X\to(-5)}1}{\lim\limits_{X\to(-5)}5X-\lim\limits_{X\to(-5)}1}$

$=\dfrac{\lim\limits_{X\to(-5)}X^2+4\lim\limits_{X\to(-5)}X-\lim\limits_{X\to(-5)}1}{5\lim\limits_{X\to(-5)}X-\lim\limits_{X\to(-5)}1}=\dfrac{(-5)^2+4(-5)-1}{5(-5)-1}=\dfrac{-2}{13}$

21. $\lim\limits_{X \to (-15)}\left(\dfrac{X^2-225}{X+15}\right) = \lim\limits_{X \to (-15)}\dfrac{(X-15)(X+15)}{X+15} = \lim\limits_{X \to (-15)}(X-15) = -30$

22. $\lim\limits_{X \to 20}\left(\dfrac{X-20}{X^2-400}\right) = \lim\limits_{X \to 20}\dfrac{X-20}{(X-20)(X+20)} = \lim\limits_{X \to 20}\dfrac{1}{X+20} = \dfrac{1}{40}$

23. $\lim\limits_{X \to (-17)}\left(\dfrac{X^2+14X-51}{X+17}\right) = \lim\limits_{X \to (-17)}\dfrac{(X+17)(X-3)}{X+17} = \lim\limits_{X \to (-17)}(X-3) = -20$

24. $\lim\limits_{X \to (-11)}\left(\dfrac{X^2+8X-33}{X^2+12X+11}\right) = \lim\limits_{X \to (-11)}\dfrac{(X+11)(X-3)}{(X+11)(X+1)} = \lim\limits_{X \to (-11)}\dfrac{X-3}{X+1} = \dfrac{-14}{-10} = \dfrac{7}{5}$

25. $\lim\limits_{X \to 5}\left(\dfrac{X^2+10X-75}{X^2+3X-40}\right) = \lim\limits_{X \to 5}\dfrac{(X-5)(X+15)}{(X-5)(X+8)} = \lim\limits_{X \to 5}\dfrac{X+15}{X+8} = \dfrac{20}{13}$

26. $\lim\limits_{X \to \infty}\left(\dfrac{7X^2+2X-4}{X^2+4X-1}\right) = \lim\limits_{X \to \infty}\dfrac{\frac{7X^2}{X^2}+\frac{2X}{X^2}-\frac{4}{X^2}}{\frac{X^2}{X^2}+\frac{4X}{X^2}-\frac{1}{X^2}} = \lim\limits_{X \to \infty}\dfrac{7+\frac{2}{X}-\frac{4}{X^2}}{1+\frac{4}{X}-\frac{1}{X^2}} = \dfrac{7}{1} = 7$

27. $\lim\limits_{X \to \infty}\left(\dfrac{3X^3-4X^2-4}{4X^4-3X^2-1}\right) = \lim\limits_{X \to \infty}\dfrac{\frac{3X^3}{X^4}-\frac{4X^2}{X^4}-\frac{4}{X^4}}{\frac{4X^4}{X^4}-\frac{3X^2}{X^4}-\frac{1}{X^4}} = \lim\limits_{X \to \infty}\dfrac{\frac{3}{X}-\frac{4}{X^2}-\frac{4}{X^4}}{4-\frac{3}{X^2}-\frac{1}{X^4}} = \dfrac{0}{4} = 0$

28. $\lim\limits_{X \to \infty}\left(\dfrac{5X^4-3X^2+7}{3X^4+2X^2-1}\right) = \lim\limits_{X \to \infty}\dfrac{\frac{5X^4}{X^4}-\frac{3X^2}{X^4}+\frac{7}{X^4}}{\frac{3X^4}{X^4}+\frac{2X^2}{X^4}-\frac{1}{X^4}} = \lim\limits_{X \to \infty}\dfrac{5-\frac{3}{X^2}+\frac{7}{X^4}}{3+\frac{2}{X^2}-\frac{1}{X^4}} = \dfrac{5}{3}$

29. $\lim\limits_{X \to \infty}\left(\dfrac{3X^2-2X+1}{12X^2+5X-1}\right) = \lim\limits_{X \to \infty}\dfrac{\frac{3X^2}{X^2}-\frac{2X}{X^2}+\frac{1}{X^2}}{\frac{12X^2}{X^2}+\frac{5X}{X^2}-\frac{1}{X^2}} = \lim\limits_{X \to \infty}\dfrac{3-\frac{2}{X}+\frac{1}{X^2}}{12+\frac{5}{X}-\frac{1}{X^2}} = \dfrac{3}{12} = \dfrac{1}{4}$

30.

$\lim\limits_{X \to \infty}\left(\dfrac{8X^3-2X^2+X+1}{7X^3-4X-1}\right) = \lim\limits_{X \to \infty}\dfrac{\frac{8X^3}{X^3}-\frac{2X^2}{X^3}+\frac{X}{X^3}+\frac{1}{X^3}}{\frac{7X^3}{X^3}-\frac{4X}{X^3}-\frac{1}{X^3}} = \lim\limits_{X \to \infty}\dfrac{8-\frac{2}{X}+\frac{1}{X^2}+\frac{1}{X^3}}{7-\frac{4}{X^2}-\frac{1}{X^3}} = \dfrac{8}{7}$

31. a). $f(7) = 2(7) - 5 = 9$

  b). $\lim_{X \to 7^-} f(X) = 9 = \lim_{X \to 7^+} f(X)$

  $\lim_{X \to 7} f(X) = 9$

  c). $f(7) = \lim_{X \to 7} f(X)$

  $f(X)$ is continuous at $X = 7$

32. $f(5) = \dfrac{3}{5 - 5} = \dfrac{3}{0} = undefined$

  $f(X)$ is not continuous at $X = 5$

33. a). $f(5) = 5^2 - 3 = 22$

  b). $\lim_{X \to 5^-} f(X) = 22 = \lim_{X \to 5^+} f(X)$

  $\lim_{X \to 5} f(X) = 22$

  c). $f(5) = \lim_{X \to 5} f(X)$

  $f(X)$ is continuous at $X = 5$

34. a). $f(5) = \dfrac{5}{5 - 1} = \dfrac{5}{4}$

  b). $\lim_{X \to 5^-} f(X) = \dfrac{5}{4} = \lim_{X \to 5^+} f(X)$

  $\lim_{X \to 5} f(X) = \dfrac{5}{4}$

  c). $f(5) = \lim_{X \to 5} f(X)$

  $f(X)$ is continuous at $X = 5$

35. a). $f(-2) = \dfrac{2(-2) + 5}{(-2)^2 + 5(-2) + 6} = \dfrac{1}{0} = undefined$

  $f(X)$ is not continuous at $X = -2$

36. a). $f(7) = \dfrac{7 - 7}{7^2 + 3(7) + 2} = \dfrac{0}{72} = 0$

  b). $\lim_{X \to 7^-} f(X) = 0 = \lim_{X \to 7^+} f(X)$ $\lim_{X \to 7} f(X) = 0$

  c). $f(7) = \lim_{X \to 7} f(X)$ $f(X)$ is continuous at $X = 7$

37. a). $f(11) = \dfrac{2(11) - 1}{(11)^2 - 15(11) + 44} = \dfrac{21}{0} = undefined$

  $f(X)$ is not continuous at $X = 11$

CPSIA information can be obtained
at www.ICGtesting.com
Printed in the USA
LVHW060926190322
713863LV00029B/802